International and Development Education

The *International and Development Education Series* focuses on the complementary areas of comparative, international, and development education. Books emphasize a number of topics ranging from key international education issues, trends, and reforms to examinations of national education systems, social theories, and development education initiatives. Local, national, regional, and global volumes (single authored and edited collections) constitute the breadth of the series and offer potential contributors a great deal of latitude based on interests and cutting edge research. The series is supported by a strong network of international scholars and development professionals who serve on the International and Development Education Advisory Board and participate in the selection and review process for manuscript development.

Institute for International Studies in Education
School of Education, University of Pittsburgh
5714 Wesley W. Posvar Hall, Pittsburgh, PA 15260

Center for International and Development Education

Graduate School of Education & Information Studies, University of California, Los Angeles
Box 951521, Moore Hall, Los Angeles, CA 90095

Titles:

Higher Education in Asia/Pacific: Quality and the Public Good
Edited by Terance W. Bigalke and Deane E. Neubauer

Affirmative Action in China and the U.S.: A Dialogue on Inequality and Minority Education
Edited by Minglang Zhou and Ann Maxwell Hill

Critical Approaches to Comparative Education: Vertical Case Studies from Africa, Europe, the Middle East, and the Americas
Edited by Frances Vavrus and Lesley Bartlett

Curriculum Studies in South Africa: Intellectual Histories & Present Circumstances
Edited by William F. Pinar

Higher Education, Policy, and the Global Competition Phenomenon
Edited by Laura M. Portnoi, Val D. Rust, and Sylvia S. Bagley

The Search for New Governance of Higher Education in Asia
Edited by Ka-Ho Mok

Forthcoming titles:

International Students and Global Mobility in Higher Education: India, China, the United States and Other Key Countries
Edited by Rajika Bhandari and Peggy Blumenthal

Policy Debates in Comparative, International, and Development Education
Edited by John N. Hawkins and W. James Jacob

Curriculum Studies in Brazil: Intellectual Histories, Present Circumstances
Edited by William F. Pinar

Increasing Effectiveness of the Community College Financial Model: A Global Perspective for the Global Economy
Edited by Stewart E. Sutin, Daniel Derrico, Rosalind Latiner Raby, and Edward J. Valeau

Access, Equity, and Capacity in Asia Pacific Higher Education
Edited by Deane Neubauer and Yoshiro Tanaka

Affirmative Action in China and the U.S.

A Dialogue on Inequality and Minority Education

Edited by

Minglang Zhou

and

Ann Maxwell Hill

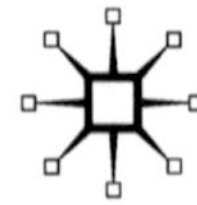

AFFIRMATIVE ACTION IN CHINA AND THE U.S.

First published in hardcover in 2009 by
PALGRAVE MACMILLAN®
in the United States—a division of St. Martin's Press LLC,
175 Fifth Avenue, New York, NY 10010.

Where this book is distributed in the UK, Europe and the rest of the world, this is by Palgrave Macmillan, a division of Macmillan Publishers Limited, registered in England, company number 785998, of Houndmills, Basingstoke, Hampshire RG21 6XS.

Palgrave Macmillan is the global academic imprint of the above companies and has companies and representatives throughout the world.

Palgrave® and Macmillan® are registered trademarks in the United States, the United Kingdom, Europe and other countries.

ISBN: 978–0–230–61334–8

Library of Congress Cataloging-in-Publication Data

Affirmative action in China and the U.S. : a dialogue on inequality and minority education / edited by Minglang Zhou and Ann Maxwell Hill.
p. cm. — (International & developmental education)
Includes bibliographical references and index.
ISBN 0–230–61235–0
1. Affirmative action programs in education—China. 2. Discrimination in education—China. 3. Minorities—Education—China. 4. China—Ethnic relations. I. Zhou, Minglang, 1954– II. Hill, Ann Maxwell.

LC213.53.C6A34 2009
379.2_60951—dc22 2009007654

A catalogue record of the book is available from the British Library.

Design by Newgen Imaging Systems (P) Ltd., Chennai, India.

First PALGRAVE MACMILLAN paperback edition: September 2010

10 9 8 7 6 5 4 3 2 1

Printed in the United States of America.

Contents

Part II Between State Education and Local Cultures

Part III Between Market Competitiveness and Cultural/Linguistic Identities

Part IV Globalizing the Discourse on Inequality and Education

Figures and Tables

Figures

Tables

Foreword

Colin P. Mackerras

There can be no doubt about the importance of the main topics discussed in this volume, namely, ethnicity and ethnic education of the countries of focus, China and the United States.

All over the world, ethnic issues have been extraordinarily controversial. Since the collapse of the Soviet Union at the end of 1991 greatly reduced the world struggle between capitalist liberal democracy and socialism, ethnic issues have come to the fore as a source of conflict. Ethnic inequalities can be found in almost all the world's countries, but there is no doubt that they can exacerbate already existing resentments, even leading to violence. Although most modern states have adopted affirmative action or preferential policies to try and reduce inequalities between majorities and minorities, ethnic conflict persists in most parts of the world. Ethnicity is a topic on which people feel very strongly and on which disrespect inevitably invites hostile reaction. Few states in the world can really claim to have handled ethnic issues well and in my opinion few are entitled to engage in finger-pointing.

Among antidotes for inequality, education holds a primary position. Of course, equality in education can rarely solve problems. But it does have the potential to ameliorate their worst effects. Education is about increasing human capital, and about enabling people to fulfill their potential and to obtain good jobs in the workforce. Failure usually results in resentment and bitterness against society, which can lead to great hostility and even open conflict. There can be few issues of greater importance for both a nation's society and economy than education.

I attended the conference in 2006 on the Dickinson College campus that resulted in the writing of this book. Although I decided against contributing a chapter, I believe the conference was of immense value. It brought out issues of very great significance, and produced a range of points of view, which is essential to first-rate scholarship. One point of great significance is that the conference, and consequently this book, included papers both about China and the United States. These two countries matter for their ethnic policies and for the theories of multiculturalism they bring out. In many ways these two countries are very different indeed in policy and reality. Yet despite frequent accusations heard in both countries about the other over ethnic matters, especially in the United States against China, this book shows that they actually share quite a few problems in common. I believe very strongly that the organizers of the conference have done a magnificent job both in taking the initiative to hold the conference in the first place and to bring this book to fruition.

One of the points I found impressive about the conference and also see replicated in this book is the number of scholars who write from basic personal experience, which their insight and command of information have enabled them to translate into excellent research. Of course this applies both to the Chinese and American participants. What we find is primary research both on China and the United States, with an excellent comparative perspective as well. The editors have given play to positive policies, but they certainly do not ignore social tensions and I admire the balanced coverage they have achieved.

Considering the disturbances that occurred in the Tibetan areas of China in the first half of 2008 and the strong divergent feelings throughout the world on the subject of Tibet, I strongly welcome the chapters on Tibetan education in this book. Among the causes of the disturbances was frustration among Tibetans born of their inability to secure employment that matched their efforts. Education is one of the main ways of addressing these problems, and it is good to see sane and well-considered treatments by people who really understand the complexities of the issues.

I commend this book and its editors and authors to a wide readership. A balanced and fair treatment of an extraordinarily important set of topics, it deserves to make an impact.

COLIN MACKERRAS
October 2008

Acknowledgments

This book began with a lively conference on the Dickinson College campus in April 2006. Our theme was affirmative action in China's education for minorities (internationally referred to as "positive policies"), and our participants converged on the campus from the People's Republic of China, Hong Kong, Australia, and various U.S. universities. We, the editors, are grateful to Dickinson College for their support for the conference and help with preparing the manuscript for this volume that has drawn on conference participants' papers. We are also grateful to the Freeman Foundation for its grant to our East Asian Studies Department, a grant that made academic activities, such as this symposium, possible.

We also thank our participants, a group that included not only scholars, but policy planners and students as well. The most significant shared attribute of over half of our participants was their direct experience—as beneficiaries, scholars, or administrators—of affirmative action programs in China and the United States. The contributors to this volume, most of whom were conference participants, inject into the volume's interrelated discussions significant minority voices, conveying their sense of urgency over the declining prospects for the education of minority students, and in many cases, for the survival of their cultures, languages, and identities.

Special thanks to Colin Mackerras for his provocative insights as one of our conference discussants, his timely review of the conference (*Asian Ethnicity* 2006: 202–306), and the foreword to this volume.

Series Editors' Introduction

John N. Hawkins and W. James Jacob

This book is part of Palgrave Macmillan's *International & Development Education Book Series*, which focuses on the complementary areas of comparative, international, and development education. Books in this series emphasize a number of topics ranging from key international education issues, trends, and reforms to examinations of national education systems, social theories, and development education initiatives. Local, national, regional, and global volumes (single authored and edited collections) constitute the breadth of the series and offer potential contributors a great deal of latitude based on interests and cutting edge research. The series is supported by a strong network of international scholars and development professionals who serve on the International & Development Education Review Board and participate in the selection and review process for manuscript development.

This edited volume by Minglang Zhou and Ann Maxwell Hill, the second to appear in the *International & Development Education* series, comprises an introduction and fourteen well-conceptualized chapters that address affirmative action, an often-contested issue in China and the United States. Chapter contributions stem from a conference held at Dickinson College in April 2006 that received international recognition for its focus on Chinese minority education issues. The editors have extensive experience in research on China's minorities and have brought together an international group of scholars, administrators, and policy makers who provide multiple case study and topical chapters on minority education issues in both China and the United States. The contributors to this volume also raise the importance of preservation of indigenous and minority education through an emphasis on minority languages, identities, and cultures.

Divided into four sections, the volume begins with an introduction by the editors that sets the stage in terms of both historical background and theoretical underpinnings for the book. The editors provide an historical overview of the development of ethnic minority status and education policies in China. While the introductory chapter and first three sections of the book (comprising chapters one–eleven) focus on Chinese ethnic minority issues and case studies, the concluding section turns to the case of the United States.

The editors and contributors refer to "China's positive policies" repeatedly throughout the book and specifically in part I on "Debating China's Positive Policies: Historical Antecedents and Contemporary Practice." Contributors identify how these positive policies relate to Chinese ethnic minorities and ethnic minority education. These positive policies constitute a shift away from the former Soviet-based model of ethnic minorities that is arguably one of the primary factors that led to the demise of the former Soviet Union. While China has moved more toward what we might call "affirmative action" and away from the "nationalities" policy, there continues to be a unwritten Hanification policy at work, especially in critical border and frontier areas. In other words, China has its own special history and cultural approach to minority education issues. There were periods (e.g., Great Leap Forward, Cultural Revolution, etc.) where social class analysis trumped ethnic diversity issues.

In part II, three chapter case studies are provided around the topic of "Between State Education and Local Cultures." Two cases studies focus on Tibet and another on ethnic minority school consolidation in rural Sichuan Province. Part III turns to how the market economy and its competitiveness intersect with ethnic minority issues of culture, language, and identity. Three additional chapter case studies are provided in this section with examples of ethnic minority groups in Shandong, Xinjiang, and Gansu Provinces.

The final section is labeled "Globalizing the Discourse on Inequality and Education" and shifts to affirmative action issues in the United States. Forty years of affirmative action history is introduced in chapters twelve and thirteen with references to key court decisions at both the federal and state levels. Only a cursory mention or comparison is given in several chapters between China and the United States until the concluding chapter that provides a more complete comparative example. In this chapter, Hill traces several similarities and differences between Native Americans and China's minority nationalities by highlighting various accommodations—not always peaceful—reached between powerful minorities and the Chinese court, including the rise of non-Han peoples such as the Mongols and Manchus to political dominance, in contrast to the unremitting pressures on and marginalization and impoverishment of many Native Americans in the United States.

This book represents a major work on Chinese minority education and is an example of how a group of international scholars can band together to address a contested issue such as affirmative action in two national education settings. The book serves as an excellent source for individuals interested in studying historical trends that identify areas of equality and significance in education as they relate to the preservation of ethnic minority education opportunities and in the preservation of their languages, cultures, and identities.

Introduction

Ann Maxwell Hill and Minglang Zhou

In multiracial and multiethnic societies, de facto inequalities always exist. More often than not, racial and ethnic minorities not only confront discrimination, they also experience disadvantages in education, employment, housing, and everyday life. "Positive" policies, more familiarly known in the West as "affirmative action" policies, have been widely adopted by modern states to redress historic inequalities among ethnic groups, to reduce the potential for ethnic conflict, and, at times, to enhance opportunities for the dominant group itself (Jaladi and Lipset 1992–93, 603). The People's Republic of China (PRC) is no exception and has, since its founding, deployed positive policies in education and employment, and, lately, in support of minority entrepreneurs.

Recent changes over the last two decades have exacerbated the inequalities that have historically challenged the PRC's minorities. For example, economic globalization has intensified competition for scarce resources, with the result that the workings of the market and society have increasingly taken precedence over those of the state. This has heightened concerns over the closely intertwined issues of equity for minorities and political stability for the state. This volume is the first to comprehensively examine China's positive policies in the critical area of minority education, the most important conduit to employment and economic success in the PRC after the economic reforms begun in the late 1970s.

In the United States, affirmative action policies in education are offered as a remedy for inequalities arising from past practices or conditions. However, American liberals and conservatives have long disagreed about the effectiveness, necessity, and fairness of such policies. In China, "positive (action) policies" in education, which are similar to affirmative action policies in the United States, though much broader in scope, have become increasingly controversial for some of the same reasons. After thirty years of economic reforms leading to the current intensification of market competition, and the concomitant withdrawal of central government funding for local education, positive policies in education for China's fifty-five minority groups have moved to the center of recent discussions among scholars, educators, and policy makers. Some of their arguments make direct reference to U.S. affirmative action and its controversies that are the focus of the final section of this volume. While not as heated as the debates in the United States over affirmative action, discussions in China nonetheless have raised new questions about the effectiveness and fairness of positive policies that directly affect over 9 percent of the population (106 million out of a total population of 1.3 billion) officially designated as minority nationalities, and indirectly impact local Han populations in the majority. And at the highest levels of PRC policy making, perennial questions about the economic development of minority regions, and minorities' overall political and legal status, have been revisited in light of increasing strains in the older, Soviet model of the multinational state that historically has served as the blueprint of the Chinese Communist Party (CCP) for nation-building.

Why these debates and why now? The responses of our contributors are multifaceted, coming at the question from experience with different national agendas, sometimes from opposite sides of the debates within the nation, and sometimes originating in firsthand research on education in minority communities. However, their discussions clearly pivot around four factors that are essential to understanding the current impetus for, and significance of, China's positive policies. The first is the legacy of China's past policies in education extending back into imperial times. Education of the empire's diverse populations played a role in the security of the empire's frontiers and political integrity, just as it did later as China's nation-builders searched for models for political unification that necessarily had to acknowledge the power of ethnic groups beyond the Han majority. The Soviet multinational state-building model was one solution to this problem.

The second factor is precisely the demise of the Soviet model in the PRC beginning in the mid-1990s, and its replacement by a new concept of the nation-state, *duo yuan yi ti* (one nation with diversity), that rhetorically resembles Western multiethnic states with their guarantees of equality for all citizens. In effect, this new model seems designed to compete with those in capitalist states for assuring justice and equality for minority groups. Yet the new model—one that demonstrates a renewed commitment to explicit positive policies in education, employment, and business—retains the administrative and jural apparatus privileging the status of the ethnic group over the individual, and maintains the political preeminence of the Han-dominated central government over minority autonomy, though it shows more flexibility than the Soviet model.

The third and fourth factors, the cumulative impact of the PRC's economic reforms since the 1980s, and the default of government to local resources in many areas of public services, especially during the administration of Jiang Zemin and Zhu Rongji (1990–2003), are closely related. The reintroduction of markets, the gist of the economic reforms, has had differing consequences for China's diverse populations, especially minorities, since the 1990s. The concomitant retreat of government from direct provision of jobs and basic services, such as education and health care, has exacerbated the negative impact of the economic reforms, widening the gap between China's eastern, coastal region and the less-developed western region, between urban and rural populations, and between the Han majority and those living in China's minority areas. The cost of education, the end of government-guaranteed employment for high school and university graduates, and the competitive, globalized, labor market that selects for those with skills, good educational credentials, and fluency in the national language (and even in English) at worst puts most minority students at a decided disadvantage. At best, this new political economy favors those from minority areas who successfully distance themselves from indigenous languages, cultures, and identities.

Understanding issues of education for China's minorities necessitates some familiarity with the complex, highly politicized terrain of ethnic classifications and the recent trajectory of positive policy evolution in China. These overviews are the backdrop for our discussion of the authors' contributions that follows. The latter are organized around the debates over positive policies, tensions between state education and minority cultures, the costs of market competitiveness for linguistic and cultural identities, and comparative views of "affirmative action" discourses in China and the United States.

Overview: The Politics of Ethnic Identification in the PRC

Officially, China has fifty-six nationalities, if one counts the Han, China's dominant majority. The term "minority," which we and most of the authors use to talk about any of the other fifty-five groups, is a translation of the term in Mandarin Chinese, *shao shu minzu* (literally, peoples with small populations relative to the Han majority). *Minzu* can mean "nationality," "race," and, more recently, "ethnic group." In the early twentieth century, the term entered the Chinese language from Japanese at a time when both nations were borrowing concepts from European discourse to talk about peoples, national identity, and the constitution of the modern nation state (Liu 1995, 292). *Minzu*, then, was a critical term in nation-building projects throughout the twentieth century for both the Chinese Nationalists and the CCP (see Gladney 2004, 14–25; 35–38), but was not in general use during China's imperial era. The current state's project for identification of nationalities began in the early 1950s. As the PRC moved forward with the process of consolidation of its territorial and administrative apparatus, the pressing need for the identification of minorities by name and location became apparent. How could the state implement minority policies, especially those based on some notion of territorial autonomy, if it did not know who the minorities were and where they lived?

In 1954, in an initial attempt to register minorities, the government recorded over 400 minorities nationwide. In southwest China's Yunnan Province alone, over 300 minorities registered with the government, which reduced that number to 132 in its report to the central government (Zhou 2003, 8–9). The problem was less the multitude of minority claims for recognition and more the lack of uniformity among standards used by local people for self-identifying as a *minzu*. The Soviets, as they had for minority policy overall, again provided the solution. Chinese ethnologists adopted Stalin's four criteria for identifying and making distinctions among minorities: common language, common territory, common economy, and common psychology (culture; see Dreyer 1976; Mackerras 1994). On the ground, this was actually a complicated project. Ethnologists were well aware that what worked in the Soviet Union, or at the level of abstraction, had to be adapted to the complex, specifically Chinese realities. For example, in frontier areas minority populations lived dispersed among the Han and were seen as having begun to assimilate to Han culture. Who was a minority and who was Han? And although officials and the people themselves were supposed to collaborate with the ethnologists in decision-making, the final determination of official *minzu* status rested with the central government. By 1964 the process of minority identification was largely complete; only two groups were added later (see Wang et al. 1998, 106–118).

The implications of a state-sanctioned classification of nationalities are profound. Western scholars since Bernard Cohn's early work on British India (1987) have maintained that states are about ordering—counting, mapping, and naming, for example—as a way of facilitating state functions. The state, then, has the capacity to create and enforce, in effect, to naturalize, new social and political realities. In this light, after half a century, the minority categories sanctioned by the PRC have predictably taken on a life of their own. They have been institutionalized through local participation in county-level political consultative conferences and legislatures in minority areas up to the level of the national assembly, not to mention through the careers of minority cadres who staff local governments in autonomous minority areas. Add to this the central government policies that exempted minorities from the stricter provisions of the national family-planning policy imposed upon the Han and gave minorities preferential treatment in access to education, among other prerogatives. Investment in state-sanctioned ethnic identities has been especially pronounced among minority elites, often bilingual and bicultural, whose status in national political and cultural arenas is virtually synonymous with their roles as representatives of their *minzu*.

Historically, the term *han* (as distinct from *han minzu*) has undergone many iterations, most designating populations in North China at the center of empire, others more ambiguous. The contemporary Han *minzu* ostensibly obviates some of the ambiguity in the denotation and scope of the earlier term, although those designated as Han *minzu* today include peoples with a wide range of cultural practices and speaking eight, almost mutually unintelligible, Chinese languages. Ironically, given the political clout of the idea of *minzu* in the context of nation-building, China's Han majority (91 percent of the population) in everyday practice is not usually thought of as *minzu* at all. Like whites in the United States, the majority Han are implicitly the standard by which all other groups are judged and usually not seen as "ethnic," one of the strongest connotations of a *minzu* identity. With the exception of cases where the Han live in areas numerically dominated by minorities, or in cities where the admissions policies of urban universities may benefit disproportionately the city's small minority populations, the Han in general are more successful in getting educated and finding employment than minorities.

Overview: Racial/Ethnic Equality amid Changing International and Domestic Politics

During the Cold War, the PRC adopted the Soviet model of multinational state-building, seeming to compete with capitalist states for the moral high ground in assuring equality for minority groups in the political sphere. As a response to the collapse of the Soviet Union in 1991, the PRC has been making efforts to replace the old Soviet model with the Chinese model of "one nation with diversity" (*duo yuan yi ti*) in handling the increased imbalances in socioeconomic equality brought about by China's embrace of economic globalization (see Zhou forthcoming). Thus, the PRC's current management of ethnic equality has to be considered in the contexts of two transitions: first, from a socialist, state-planned economy to a "capitalist, free-market" economy, and second, from the Soviet model of multinational state-building to the Chinese model of "one nation with diversity." Though still labeled "socialist," the PRC's current economic and nation-state models now share a number of similarities, at least on the surface, with those in the United States. China also shares with the United States some of the same dilemmas in seeking racial/ethnic equality.

As we have noted, for the past two decades, the PRC's positive policies have come under increasing criticism, particularly in the area of education.

Why have these questions arisen in China during this time? How will China respond to them? How has China addressed the failure of the Soviet model of multinational state-building? What has China done to maintain ethnic equality in the ferocious process of marketization? How have minorities fared in education during the economic and political transitions?

Answers to these questions have significant implications for understanding the changing context of minority autonomy and issues of national integration. In a larger sense, they also have implications for policy formation and implementation in China's emerging market economy and transition to the "one nation-state with diversity" during the process of globalization.

The contributors to this volume try to answer these questions from both international and domestic perspectives, as well as from their research, and sometimes home experience, in one of China's many minority communities. The following analysis of macro changes in China's model for ethnic relations and their concomitant policies is intended to help readers better appreciate the points made by the contributors to this volume, many of whom assume familiarity with the Chinese context of minority education.

The Adoption of the Soviet Model

As Walker Connor's chapter in this volume makes clear, the Soviet model of multinational state-building is considered the blueprint for the PRC's handling of the "national question"—a Leninist and Stalinist term for managing ethnic relations (see Connor 1984; Dreyer 1976, 2006). The Soviet Constitution provided the theoretical foundation and relevant language for rights embedded in PRC's minority policies. Three fundamental views from Marxism, Leninism, and Stalinism have theoretically guided the CCP's approach to the national question. First, as early as 1922, during its second national congress, the CCP acknowledged Lenin's view that "different nations are advancing in the same historical direction, but by very different zigzags and bypaths," and that some nations are more cultured/advanced than others (China 1981, 5; Lenin 1967, 172). According to this view, being culturally and economically different, China's minorities had the right to be politically different, that is, to have the right to self-determination during the Republican period and to territorial autonomy during the PRC, so they could eventually catch up with the "advanced" Han in economic and cultural development. Moreover, this view also underpins the Han big-brother-style assistance to minorities in their economic and cultural development, and the positive policies during the PRC.

Second, also during the 1922 congress, the CCP adopted the Leninist categorization of nations into oppressors and oppressed, and advocated fighting against the oppressors and winning true equality among nations as an integral part of the overall communist revolution (see China 1981, 1–2; Lenin 1967, 749). However, not until 1923 at the CCP's third congress did it explicitly put forward the principle of national equality in China where the idea of confederation was seen to play a significant role (see China 1981, 5; Wen et al. 2001, 11). This principle is later enshrined in the PRC's provisional constitution, the Common Program of the Chinese People's Political Consultative Congress (General Principles), and the 1954 Constitution (General Principles) and its later versions. Third, while acknowledging early on Lenin's view on the historical development of nations, the CCP did not emphasize until the 1950s the Stalinist view that "nation" is a historical category and nations undergo three stages: formation, conflict, and convergence (see Stalin 1975, 153–156; Wen et al. 2001, 147–148). This view assumes that nations will converge during the communist rule in the Soviet Union, China, and elsewhere. Thus, the national question would be naturally resolved during socialism. In short, these three essential Marxist-Leninist-Stalinist views together underlined the CCP's theoretical considerations of the national question in China and were the foundations for the CCP's policies and the PRC's laws until the 1990s.

In practice, China modeled its laws and policies after those of the Soviet Union and received advice from the Soviet leadership and advisers (Zhou 2003, forthcoming). Before the PRC was established, the Chinese Soviet passed its constitution in 1931, a constitution clearly modeled on the 1918 Russian Soviet Constitution (see China 1981, 16–17; 1987, 10–11). The Chinese Soviet Constitution stipulated that, regardless of their nationalities, all members of the working class had the right to vote for, and be elected to, government offices, had freedom of religion, and had the right to self-determination or autonomy. The essence of these stipulations is found in the 1918 Constitution of the Russian Soviet Federated Socialist Republic, which contained the rights to self-determination, freedom of religion, and participation in elections (see Strong 1937).

In drafting the 1954 constitution of the PRC, China essentially continued the same rights for minorities while working in close consultation with the Soviet Union. First, the PRC leadership and the Soviet leadership had communication on minority political rights in the new PRC. It is reported that, in early 1949, during his visit to Mao Zedong (chairman of the CCP, 1945–76) in China, A. I. Mikoyan passed Stalin's message that the CCP should not allow minority self-determination but give minority groups only territorial autonomy (Ledovskii 1999/2001, 85). Though Mao Zedong had the same idea as early as 1936 (see Liu 1996, 305–306), Stalin's advice might have influenced how the PRC's provisional constitution stayed away from the notion of self-determination, a notion that Mao Zedong and Li Weihan (a CCP leader in charge of ethnic affairs between 1948 and 1962) decided to exclude during preparations for passage of this constitution (Luo 2001, 961). Immediately after the passage of this constitution, on October 5, 1949, the CCP Central Committee instructed its regional bureaus and field-army CCP committees that the term "self-determination" should no longer be used in its minority policy because imperialists and minority reactionaries might use it to sabotage the unification of China (China 1997, vol. 1, 24–25).

Second, Soviet advisers examined the draft of the PRC's 1954 Constitution and suggested many revisions. After revising the document, China sent the draft to Moscow for comments before the People's Congress formally passed it in September 1954. Essentially China adopted the Soviet model by constitutionally endorsing three basic principles: equality for all national minorities, territorial autonomy, and equality for all minority/national languages and cultures.

Who should enjoy these constitutional rights? From 1949 to 1979, as we noted earlier, the PRC went through a process of identifying and classifying various minority communities into fifty-five groups on the basis of Stalin's (1975, 22) criteria of a common language, territory, economic life, and culture. Only the officially recognized groups were entitled to the constitutional right to organize their own autonomous governments at the *xiang* (a subcounty administration), county, prefecture and/or regional levels, depending on the sizes of their populations and territories. The autonomous governments, which were headed and mostly staffed by members from the community in the name of which the autonomy was given, enjoyed tax reduction and financial subsidies, the right to use native languages in government and education and to develop local cultures, advantages in access to education, and flexibility in implementing the central government's policies. It is important to note that these rights and benefits were constitutionally guaranteed to those minority groups, and generally not to their individual members.

In Search of a Chinese Model

When the "mighty" Soviet Union collapsed, witnessed on TV screens around the world in 1992, the PRC government was shocked, particularly by the role that ethnic relations played in the Soviet downfall (see China 2002, 10, 37–38; Li 1999, 6). China's first reaction was to evaluate what role the Soviet model of multinational state-building had played in the fall. When its evaluation showed the model had a very negative influence, the Chinese government began to seek a replacement for the failed Soviet model in handling China's national question.

Immediately after the collapse, the Chinese government sponsored a series of studies of the relationship between ethnic relations in the Soviet

Union and its fall. These studies suggested four key lessons regarding the failure of the Soviet model of multinational state-building (Guo 1997; Mao 2001, 113–119; Tiemuer and Liu 2002, 136). First, the Soviet confederation system gave the minority republics too much political power insofar as they could legally separate from the union when they had conflicts with the union (Guo 1997, 79–80). Second, the Soviet legal system was not modernized to prevent any legitimate secession of the republics while the Soviet Union failed to rule by law (Guo 1997, 228–229; Hua and Chen 2002, 186–187). Third, the Soviet nativization (*minzuhua*) process promoted too many minority officials locally and resulted in the minority dominance of both the government organs and party apparatus in the republics (Guo 1997, 156–157). This situation allowed nationalist minority officials to overrun the communist party apparatus, costing the Soviet Union's control of the republics. Fourth, the failed Soviet economy engendered too many economic disparities that deeply separated the republics in terms of wealth and economic opportunities (Guo 1997, 129–132). These lessons suggested to the PRC government that the Soviet model might fail in China, too, if it was not updated to handle changing ethnic relations.

Fortunately for China, two years before the Soviet demise in 1989, Fei Xiaotong, a well-known anthropologist, delivered a crucial speech at Hong Kong Chinese University. In this speech, he proposed several interrelated ideas that subtly shifted the position of minorities vis-à-vis the state, all of which Fei claimed had been taking shape in his mind since the 1950s (Fei 1991; 1999, 13). Noting that the Chinese nation includes the fifty-five minority groups and the Han majority, Fei said that these groups were not simply a diverse assemblage, but a national entity that had developed from a common yearning for a shared destiny. Thus, the highest level of identity for all China's peoples is the nation, the inclusive Chinese nation (*Zhonghua minzu*). All Chinese citizens have a Chinese national identity; they are all members of the inclusive Chinese nation. Beneath this national identity are the official ethnic identities of China's fifty-six *minzu* (see Fei 1999, 13). In Fei's view of Chinese history, the Han played the core role of integrating various national elements into the Chinese nation, which has since transcended the Han to embrace the diversity of numerous ethnic groups. The concept of the diversity-in-unity of the Chinese nation assumes that the two levels of identity, national and ethnic, do not replace each other, nor contradict each other, but coexist and codevelop along with linguistic and cultural diversity. This set of interrelated ideas serves as the foundation for the contemporary model of "the Chinese nation with diversity" (*Zhonghua minzu duo yuan yi ti*), a new template for political integration in the PRC.

Fei's theory of one nation with diversity soon drew the attention of the PRC government that was in urgent need of a solution to the national question—a solution that could augment or replace the Soviet model of multinational state-building in China's political reform. The State Commission on Nationalities Affairs sponsored a symposium on this theory in May 1990, at which Fei's theory was supplemented with studies on various aspects of ethnic relations in China (see Fei 1991). In August of the same year, Jiang Zemin, president of the PRC and general secretary of the CCP (1989–2002), adopted this theory in a speech to local leaders in Xinjiang (Li 1999, 158 and 228). Jiang told the audience that the Chinese nation is made up of fifty-six ethnic groups and, in this big family of the motherland, the ethnic groups enjoy a new socialist relationship to the Han and to one another: The Han cannot do without the minorities, the minorities cannot do without the Han, and the minorities cannot do without each other. The three "cannots" have formed the CCP's basic view on the national question since then (China 2002, 18; Tiemuer and Liu 2002, 4–5). At the CCP's Sixteenth National Congress in 2002, it was further stressed that China should constantly work to increase the nation's cohesion (*ningjuli*), which Hu Jintao, the new CCP general secretary, defined as common prosperity and common economic development for every ethnic group within China (Mao and Liu 2004, 14).

The Impact of the Chinese Model on Positive Policies

The replacement of the Soviet model with the Chinese model has produced two direct consequences in the last decade. The first is the rectification of names (*zheng ming*) so they are politically correct, and the second is the revision of relevant laws to represent the essence of the new Chinese model of nation-state building.

The rectification of names has been both a philosophical and a practical issue in Chinese culture and politics since the time of Confucius. According to Confucius, "If names are not rectified, then language will not be appropriate, and if language is not appropriate, affairs will not be successfully accomplished" (trans. from Theodore de Bary and Bloom 1989, 56). Confucius pointed out the importance of discourse control and political correctness if one needs to get something done. This proved to be true for the PRC government. With the Soviet model on its way out, and the Chinese model of "one nation with diversity" newly adopted, the term *minzu* in Chinese was now given a new meaning, "ethnic identity," but the official English translation of it as nation and nationality are still tainted by the Stalinist connotations. In Beijing in 1997, the State Commission on Nationalities Affairs held a forum on whether *minzu* should be officially translated into "nation/nationality" or "ethnic group" (personal communication with scholars in Beijing, 2004). At the close of the forum, the participating experts unanimously agreed on the English term "ethnic group" or "ethnic affairs" for *minzu* because the new term can better represent the spirit of China's new discourse on minority affairs, an orientation that departs from Stalin's formulation of the national question. In the following year, the State Commission on Nationalities Affairs officially changed its English name to the "State Commission on Ethnic Affairs," or "State Ethnic Affairs Commission." The name rectification is undoubtedly to conform to China's change of nation-state model.

The replacement of the Soviet model with the new Chinese model had, and continues to have, a direct impact on China's minority policies. To address the four serious lessons from the demise of the former Soviet Union, China began to reformulate its strategy in dealing with the national question and to revise its laws on autonomy for minority areas. The new strategy, proposed in the late 1990s, was "speed up economic development but downplay the national question" (*Jiakuai jingji fazhuan, danhua minzu wenti*). The first clause in the quoted passage has been seen in China's newspapers almost everyday, but the second was passed only orally among various levels of CCP and government officials. This strategy treats rapid economic development in minority communities as the focus of China's minorities policy, because the gap between the developing Eastern coastal regions, where the Han mostly live, and the economically underdeveloped Western regions, where minorities generally live, is considered the key cause of ethnic conflicts in China currently (China 2002, 41–44 and 126–146). Therefore, in 2000, the central government launched its "Open Up the West Campaign" (*xibu da kaifei*), which aims at economically integrating Western and coastal China. If it is successful, this project is expected to eliminate or reduce the economic disparity between developed Han communities in the coastal provinces and developing minority communities in the inland provinces, and to increase the mobility between people in Han communities and those in minority communities. This project, ultimately, is intended to lead to greater ethnic, economic, geographic, and linguistic integration between the Han and the minorities and among the minorities themselves as required by the ideal of "one nation with diversity."

To curtail the political power of minority groups and to prevent any chance of legitimate power seizure by minority groups in autonomous regions, work had started in the early 1990s to examine and to revise the PRC's *Law on Minority Regional Autonomy*, which was formulated as administrative regulations in the early 1950s and passed as law in 1984. It was suggested that, since 1984, there had been two major problems in the implementation of this law, which were very similar to lessons learned from the failure of the former Soviet Union and the Eastern block (see

Mao 2001, 344–361). First, in creating local legislation for autonomy, minority groups tried to negotiate with the central government for more power to control the local economy—more power than the PRC Constitution and *Law on Autonomy* would allow. For example, minority groups wanted to specify in local legislation how much of the product and profits produced by businesses controlled by the central government should go to the autonomous governments of the regions where these businesses were physically located. Second, and more seriously, minority groups wanted to legislate more political power locally than allowed by the PRC Constitution and *Law on Autonomy.* Some minority regions, for instance, wanted to legislate the minimum percentage of positions in the local governments and legislatures that had to be filled by members of their ethnic communities, a move not allowed under the PRC Constitution and *Law on Autonomy.* Some minority regions wanted more economic power, and some wanted more political power, while many desired both. This attempt to negotiate over economic and political power was a serious challenge to the state's authority. Drawing on lessons from the Soviet failure, the PRC realized that these problems must be satisfactorily resolved before they got out of control. Also, based on the Soviet lessons as the PRC understands them, the central government realized that while it should not relinquish political power or allow nativization of the party apparatus, it nonetheless must make some concessions in the economic sphere.

On February 28, 2001, the Chinese People's Congress passed the revised PRC *Law on Autonomy for Minority Regions* (Wang and Chen 2001). This revision involved thirty-one articles of the law, including rewriting twenty, deleting two old ones, and adding nine new ones. The majority of the revised articles dealt with social and economic development, some revised articles concerned the central government's responsibilities and the proportion of officials of minority origin in autonomous governments, and three revised articles covered language use. Generally speaking, the revised law on autonomy gave local autonomous governments more power or responsibility in social and economic development, but took away some political power. For example, Articles 17 and 18 of the 1984 version of the law stipulated that the staffs of an autonomous government and its organs should include as many officials of minority origin as possible, but, in the 2001 version, these two articles require only a reasonable number of officials of minority origin on the staffs of these agencies. Clearly, this is a legal step aimed at the prevention of runaway situations like those that happened in the Soviet republics. As for economic concessions, Articles 31, 32, 34, 55, and 57, for example, gave autonomous regions more rights in foreign trade, local taxes, local budgets, and local finance, all of which are consistent with China's economic reforms.

Articles 37 and 49 of this new version of the law on autonomy, together with the PRC's 2001 *CommonLanguage Law,* have also downplayed the role of minority languages and cultures while promoting *Putonghua* as the super-language in a structured linguistic order (for more details, see Zhou 2006). For example, the requirements for the teaching of Mandarin Chinese were changed, so that it started in the early rather than the late years of education in elementary schools in minority autonomous areas (Article 37). Minority officials are now required to learn to use both standard oral Chinese (*Putonghua*) and standard written Chinese (Article 49). These measures are representative of the demotion of "nationalities" to "ethnic groups" in the new model of "one nation with diversity," where the national language is to dominate the linguistic sphere.

However, the new version of the law extends the scope of positive policies in education, employment, and business opportunities. First, in education, the new law covers minority school budgets, diversified schooling approaches, lowered admission standards for minority students, increased scholarship awards for them, and more career opportunities for minority students after graduation (Articles 37 and 71). Second, the law requires autonomous governments to give priority to minority applicants when hiring government workers and simplifies the hiring process (Articles 22 and 23). Third, it requires more tax reductions, financial and technical support, and cross-regional/border opportunities for minority businesses (Articles 57, 58, 60, 63, and 65). Finally, the revised autonomy law goes beyond positive principles by specifying some practical approaches. In May 2005, the PRC State Council, the central government, passed regulations on how to implement the law on autonomy (China 2005a). The regulations make the law administratively and financially operational in the state bureaucratic system.

However, some critical questions remain. Has the PRC taken the right approach to ethnic equality during globalization? For example, has the revision of the autonomy law done enough to address ethnic equality issues, particularly in education, during the transition from the socialist, planned economy to a capitalist, market economy? How are positive policies actually being implemented during this period? Are these policies still effective in creating or maintaining ethnic equality in a rapidly changing China?

Debating China's Positive Policies: Historical Antecedents and Contemporary Practice

Taking a broad perspective on continuities among the Soviet Union and its client states, Walker Connor in chapter one links China's minority policies, including those in education, directly to principles articulated in the Soviet-dominated Comintern during the 1920s and 1930s. Connor sees policy vacillation at the center of CCP decision-making—vacillation between minority autonomy and centralization, and between assimilation and accommodation—as responses to, and sometimes negation of, broadly Marxist principles expediently adapted by CCP leaders. Connor's perspective, representing the conventional wisdom among Western scholars whose research has focused on the PRC's minority education policies, is complemented, if not challenged, by Minglang Zhou in chapter two on the history of China's positive policies. While acknowledging that the PRC's policies in minority education reflect the influence of more general principles of Soviet policies toward minorities, Zhou can find no evidence for the role of specific Soviet advisors, or references to Soviet documents directly linked to the PRC's positive policies in minority education.

One general principle in the Soviet legacy that has played a part in China's education policies is the theoretical equality of all nationalities (*minzu*). Tiezhi Wang in chapter three takes this principle as his starting point and follows with a defense of the current PRC policy of lowering benchmark admissions scores for minority students on the national college admission examinations. His argument resonates with chapter twelve by Evelyn Hu-DeHart whose phrase, "to treat people equally, we must treat them differently," summing up her stance on affirmative action policies in the United States, captures the gist of Wang's view. Wang cites factors in minority areas, such as poverty, poor schools, and illiterate parents, that cause minority students to lag behind their Han counterparts. He argues that lowering benchmark scores for minority students taking the national college admission exams, or adding points to their scores, are appropriate measures to restore fairness, compensate for the cultural tilt of the test toward Han students, and increase the likelihood that minority students have a shot at attending some of China's best universities. In chapter four, Xing Teng and Xiaoyi Ma , both ethnologists with firsthand knowledge of the conditions in rural minority areas, point out that there are regional inequalities in education and economic development all across China that affect everyone, regardless of ethnicity. So from their perspective, Han students in areas at or below the government's official poverty level may be just as disadvantaged by the system as the minority students, yet the Han cannot avail themselves of the ethnicity-based remedies of preparatory classes and bonus points. The current system, they say, needs fixing. Teng and Ma advocate an affirmative-action system more like the one in the United States. They have in mind a system that can address the barriers to education for the individual citizen on a case-by-case basis, rather than one regulating fair access to education based only on government-sanctioned ethnic group affiliation.

How flexible can local or provincial governments actually be in the implementation of positive policies to deal with some of the "reverse discrimination" built into the central government's mandates? To answer this question, in chapter five, Yanchun Dai and Changjiang Xu, both policy practitioners turned doctoral students at Yunnan University, examine the evolution of positive policy making and implementation in Yunnan. Yunnan is the most culturally and linguistically diverse province in China. The coauthors provide numerous examples of the extent and flexibility of Yunnan's implementation of the central government's positive policies, particularly since 2000. Their analysis suggests that the abandonment of the Soviet model has given China some flexibility in addressing Teng and Ma's concerns about reverse discrimination. But their analysis also reveals serious incongruity among various positive policies at local levels and between the local and national levels. Largely due to continuing reforms, this policy incongruity has far-reaching implications for minority education, including issues of funding, teacher training, medium of instruction, student retention and graduation rates, career prospects, and even cultural and linguistic diversity itself.

Boarding Schools or Local Schools? Between State Education and Minority Cultures

As part of its modern nation-state building, in 1993, the PRC government laid out three goals for the state education system: By the year 2000, compulsory nine-year education should be implemented for 85 percent of China's total population; five–six-year compulsory elementary education should be implemented in areas at the poverty level where the remaining 15 percent of the population lived; and opportunities for a college education should be provided to 8 percent of the population between ages 18 and 21 (China 1998, 39–50). Responding to the "Open Up the West Campaign," in 2001 the PRC government reaffirmed that compulsory nine-year education should be promoted in poverty-level areas by 2010 (China 2005b). While this approach to education is fully consistent with China's rapid modernization, it is questionable whether it gives adequate consideration to local conditions.

Teng and Ma's case in chapter four for the importance of local conditions and their unintended consequences for the Han should not be interpreted as a refutation of the widely recognized fact that minority regions, areas, localities, and communities tend to be characterized more often by lower incomes, worse infrastructure, and substandard schools than those that are predominately Han. While these disparities have something to do with the proportionately greater number of minority populations found in the less-developed Western region in China, there are other factors at work.

One of the most obvious to anyone familiar with rural, minority communities is the devolution of government fiscal responsibility for public education from the central government to the locality (see chapter five for a critique of this policy). As Chan and Harrell note in chapter eight on school consolidation in a county within the Liangshan Yi Autonomous Prefecture, fiscal decentralization nationwide has meant in practice that funding for primary and secondary education in rural areas must come from local resources, such as taxes and local government enterprises. Local governments in turn have defaulted to students' families to cover the cost of village education. Cash-poor schools have resorted to various ad hoc fees to cover building expenses, schools fees, and teachers' salaries. While in Yunnan Province, as Dai and Xu note in chapter five, the provincial government has recently stepped in to eliminate some of these school fees, minority families relying on subsistence agriculture or herding often have no recourse but to keep their children home because the labor of all family members is critical to their survival. Parents' attitudes toward basic education, that is, the compulsory nine years of education mandated by law in 1986, have also changed. In chapter six on the state of basic education among Tibetan nomads, Gelek, on the basis of fieldwork conducted in 2003, found that parents are less likely than before to invest in education for their children now that the state no longer guarantees them a job, as it had in the era of the planned economy before 1980. His study demonstrates why there is concern among educators and administrators for students in remote, rural areas. Most will never reach the level necessary to sit for the national college admission exams, the target of many of China's positive policies.

Boarding at primary schools has become a necessity in mountainous minority homelands for several reasons. As Chan and Harrell in chapter eight report for the Liangshan Yi areas of Sichuan, local primary schools are closing as they are compelled to rely increasingly on resources from an impoverished local population for funding, a situation related to the decision of local governments to consolidate schools for the sake of efficiency. Gelek's work in a Tibetan autonomous prefecture in Gansu province suggests that boarding schools, recently introduced, have been the only option to serve nomadic students in sparsely populated, remote areas. Boarding schools have been favored in recent county-level policies, just as they have in Liangshan, over the development of local, rural primary schools. The contrast is stark between boarding schools in Tibetan areas for basic education and *neidi*, or inland, boarding schools, analyzed in chapter seven by Gerard Postiglione, Ben Jiao, and Ngawang Tsering. Inland schools for Tibetans, secondary schools in Han-dominated cities, have as their goal the training of minority cadres for work in their home territories and the larger aim of building national unity. As the authors explain, the 1985 policy establishing inland schools, for which there has been growing demand among Tibetans, has come to embody the PRC's reformulation of what it means to be *minzu* in the new millennium. This new model, explained in detail in this introduction, acknowledges the cultural distinctiveness of China's *minzu*, now translated as ethnic groups, while promoting an inclusive sense of the Chinese nation and identity (*Zhonghua minzu*). Based on interviews with graduates of inland schools, as well as their knowledge of the schools' structure and curriculum, the authors conclude that Tibetan students are brought into the Chinese mainstream through their inland education, but with their ethnic identity strengthened rather than diminished.

However, while the authors argue that inland boarding schools have not assimilated their Tibetan students to Han culture, most students return to Tibet with the distinct feeling that they are no longer as culturally or linguistically fluent in their home environment as before. Moreover, some have absorbed the government's view of religion as unscientific and no longer believe in Tibetan Buddhism, while others find aspects of Tibetan Buddhism "superstitious." Such trade-offs are probably inevitable and necessary if, as seems to be the case, these students are to make successful transitions to government work back in Tibet. As Zhou notes in chapter two, cadre training was an institution important to both the Nationalists and Communists, and special provisions for training minority cadres were first implemented in the CCP base area in Yan'an during the late 1930s. The current popularity of *neidi* boarding schools among Tibetans demonstrates that the objections of Tibetan parents to boarding schools evidenced in Gelek's study become moot when the boarding school educational structure is centrally planned, subsidized, and leads to a job. In the absence of sufficient data, one can only speculate whether concerns about assimilation and concomitant loss of the home culture also figure into parental calculations.

Between Market Competitiveness and Cultural/Linguistic Identities

Do a market economy and economic globalization always favor the state and educational standardization over cultural and linguistic diversity? Xiaoyi Ma, Rong Ma , and Zhanlong Ba's chapters attempt to answer this significant question from very different perspectives.

Whether it is "socialist" or capitalist, a market economy requires efficiency and competitiveness or, more plainly, the ability to make money. Both the state and minority families respond to this pressure, though in different ways. In this light, Xiaoyi Ma's discussion in chapter nine of Hui indifference to education beyond primary school is revealing. The Hui community she studied is situated in Shandong Province and dominated by what she calls "a strong business culture." Unlike Hui communities in China's northwest, the Chaocheng Hui people have much in common with the surrounding Han and have not invested much in Islamic education. Instead, Hui children attend local public schools along with Han students, but tend to drop out during middle school. They easily find work in family enterprises, and, in fact, the Hui community has higher incomes than those of the neighboring Han. Ma's study also indicates that Hui parents often see the government school teachers (Han) as discriminating against their children, another reason, in addition to their easy absorption into family businesses, for allowing Hui children to withdraw from formal education. In contrast to Tibetans, especially those in remote areas, the Hui community in Shandong has convenient access to public education, but like the Tibetan parents, Hui parents may not see the point of keeping their children in school when the economic rewards for their investment, which would include loss of labor at home, are not forthcoming. One obvious conclusion to be drawn from this similarity is that utilitarian calculations in investments in education are particularly paramount to minorities, or, reflecting more deeply on positive policies, that policies segregating minorities from Han, and subsidizing fees in secondary education, such as those characteristic of *neidi* schools, are more effective in producing mainstream attitudes and school completion than simple access to local public schooling. A related observation is that local public schools where minorities and Han are in the same classrooms may work against the kind of acculturation achieved by the *neidi* schools because local schools have minimal impact on the home cultures and reflect local ethnic politics. Discrimination in the classroom, real or perceived, may also work to discourage minority students from staying in school, as Xiaoyi Ma has noted. Nevertheless, Ma's chapter shows that sometimes the market economy may favor certain cultural traditions over others, in the face of the onslaught of state standardized education.

However, Rong Ma, in his research on bilingual education and its outcomes in Xinjiang Uygur Autonomous Region, presented in chapter ten, reports that minority graduates of regular schools in Xinjiang, where the language of instruction is *Putonghua,* are more competitive in the region's labor market than those who graduate from minority schools. Because there are fewer government jobs than graduates seeking such positions, many graduates look for work in a private sector dominated by Han managers with Han clientele and where good *Putonghua* is essential. Graduates of minority schools at the secondary level where *Putonghua* is taught as a second language do not do well in the private-sector economy, a problem beginning with the low level of their teachers' competence in *Putonghua*. Rong Ma's larger argument, based on his research, is that minority students in Xinjiang's universities are ill-served by lower admissions scores that allow them to continue their educations when their Mandarin language skills are inadequate for local employers' needs and their educational preparation puts them at a disadvantage in university classes. Furthermore, Ma says that because of positive policies favoring minority applicants to both secondary and post-secondary schools, employers have picked up the impression that minority students as a whole are not as well educated as Han students. Ma also takes note of a concern even among minority students and teachers that mandated lower admissions scores in China's positive policies encourage students to slack off and actually contribute to an increase in the gap between minority and Han students.

The question of the progress of Xinjiang minorities in education and the job market is especially acute because of Xinjiang's strategic significance to China's political integrity and security (Becquelin 2004, 359; Sautman 1998, 87). Bordered by Kazakhstan, Kyrgyzstan, Tajikistan, India, Afghanistan, Pakistan, Russia, and Mongolia, the Xinjiang Uygur Autonomous Region has a large Islamic population reflecting the proportion of Uygurs, who are Sunni Muslims, in the region's total population, about 45 percent (Bhattacharji 2008). In addition to long-standing strategic concerns, the government also must attend to the area's potential for Muslim separatist movements, the rise of unstable new states in Central Asia since the breakup of the Soviet Union, and Xinjiang's large reserves of gas and oil (Becquelin 2004, 365). In other words, Xinjiang is a region where the Chinese state has a strong stake in the success of positive policies. Thus, we might view it as a test case for the potential effectiveness of positive policies across China. Rong Ma's chapter indicates that Xinjiang's secondary schools and universities have produced too many people who are unemployable in the private sector because of lack of proficiency in *Putonghua,* and because of the widespread impression among local entrepreneurs that minority graduates are not as well qualified as Han students. So from Ma's point of view, positive policies are problematic and may even work against minority achievement. Therefore, he supports the new policy put in place by the Xinjiang government in 2004, which requires all or the majority of university-level classes to be taught in Mandarin. Ma is frank about the problems with this policy and the resistance to it, but points out that it at least functions as an early reality check for students before they go on the job market. In the view of some Uygur educational professionals at Xinjiang University, continued support for classes in minority languages and cultures at the university is important even in the midst of a shifting economy and an open labor market privileging *Putonghua* and essentially Han business skills. Clearly, the market economy and economic globalization challenge Uygur cultural and linguistic dominance in Xinjiang at the same time as the Chinese model of one nation with diversity marginalizes locally dominant minority cultures and languages nationwide.

Of more concern is whether linguistic diversity may be maintained at all with this Chinese model of nation-state building. In some circumstances the teaching of indigenous languages appears to be increasingly unpopular and not supported by the people who are, or once were, its native speakers. Zhanlong Ba in chapter eleven discusses this phenomenon in two Yugur (distinct from Uygur mentioned earlier) communities in Gansu Province. In the area he studied, only about half the total population of people officially identified as Yugur speak one of the Yugur languages, and the percentage of monolingual speakers of the local languages is falling. Ba attributes much of the failure in the two local experiments in teaching Yugur to the lack of coherence within the government's bureaucratic hierarchy, such that local decision-making is haphazard and policy implementation, unpredictable. Yet it is also clear from his ethnographic account that local people, whose school system has achieved some notable, nationwide recognition for success in basic education, are ambivalent about teaching Yugur. Some are hostile enough to embrace a language ideology that sees the Yugur languages as useless and uncultured.

Globalizing the Discourse on Racial/Ethnic Equality

The final section of this volume refocuses our attention on larger historical trends in China by contrasting them with affirmative action and its antecedents in the United States. In chapter fourteen, anthropologist Ann Maxwell Hill takes a broad, comparative overview of native peoples and assimilation in modern China and the United States with the goal of providing a fuller context for understanding each nation's affirmative action policies. With its focus on cultural, demographic, and historical factors, Hill's chapter discusses the reasons for greater assimilation pressures on Native Americans than on China's non-Han peoples.

In chapter twelve, Evelyn Hu-DeHart opens her defense of U.S. affirmative action policies first by placing them within a global perspective relative to similar policies and then particularizing them within the historical circumstances of the United States. She argues that affirmative action measures are necessary to remediate inequality, even though, as in China, race-based discrimination is against the law in the United States. Antidiscrimination laws are not enough, says Hu-DeHart, who points to widely known research showing the predominance of white males in top jobs in industry and government in the United States. She also mentions the "glass ceiling" in identifying invisible barriers that block the advancement of Asian Americans. Hu-DeHart's combative stance and attention to racial politics have few parallels in the debate over affirmative action in China. However, her attention to the workings of the job market and the problems encountered in pre-college education for minorities in the United States resonates with many conditions reported by her Chinese counterparts, who often make arguments for the Chinese cases, citing the U.S. affirmative action policies, without a comprehensive understanding of its history and contexts. Douglas E. Edlin, in chapter thirteen, follows up Hu-DeHart with a review of the legal history of affirmative action policies in the United States, starting with DeTocqueville's well-known observation that "scarcely any political question arises in the United States which is not resolved, sooner or later, into a judicial question." Edlin fleshes out and contextualizes many of the cases mentioned by China-side contributors in chapters three and four, at the same time providing a clear exposition of the origins of affirmative action in U.S. education. His focus, like that of many of our contributors, is primarily on the university admissions process.

Conclusions

Our purpose in this introduction has been to identify reasons for the urgency and salience of debates in China over its positive policies in minority education in the new millennium, when positive policies have long been a mainstay of the modern Chinese state and can be found even earlier in imperial practices. Although still nominally socialist and under the authority of the CCP, Chinese society manifests the cumulative effects of more than thirty years of reforms, economic and political, that by 2008 have produced the world's second largest economy. Clearly, the very success of these reforms, predominantly in China's eastern cities and coastal areas, and the dismantling of the central government's role in the provisioning of education, employment, and other socialist supports of the pre-1990 planned economy, have introduced new stresses for China's minority populations, most of them already disadvantaged in the burgeoning, capitalist, market economy by geography or poverty or language or access to education, or some combination of these conditions.

To some extent, recent policy changes affecting minority education can be seen as directly related to the changes in China's economy and state concerns that uneven development poses a serious threat to national integration. As we have shown, China's revised *Law on Autonomy for Minority Regions* (2001) extends the scope of the earlier law, mandating positive policies that explicitly provide more support for minorities' access to education, of which the most controversial, as our contributors have indicated, is the requirement that schools accept lower admission scores for minority applicants than the Han students on entrance exams. In the same year, the PRC passed the *Common Language Law*, which puts state authority and resources behind the promotion of *Putonghua*. Minority language rights are explicitly protected in the same law, but the contradiction represented by the *Common Language Law* pits local, often poorly resourced, languages against a national language, the obvious utility of which, in the closely linked education and economic spheres, tilts the long-term outcome in favor of *Putonghua*. Not surprisingly, several of our chapters identify minority communities where the choices they articulate support the use of *Putonghua* over local languages as the medium of instruction in schools.

It may be mistaken, then, to evaluate the effectiveness of the PRC's positive policies in minority education simply on the basis of their success in giving minorities more access to good education and jobs. Assessing the PRC's positive policies also means examining their impact on the assimilation of minorities to the dominant Han culture and to the use of *Putonghua*, the "standard Chinese national language." The new model for national integration, *duo yuan yi ti*, officially promotes and privileges an overarching national identity, departing from the earlier model with its emphasis on equality and concomitant autonomy among *minzu*, and buttressing the inevitability of other "standardizations," such as standardized textbooks and curriculum, in a nation where the Han make up more than 90 percent of the population. Eventually standard language and standard education may contribute significantly to the cohesion of the inclusive Chinese nation, only if positive policies can help to advance minorities up the educational and socioeconomic ladders in the Chinese mainstream.

References

Becquelin, Nicholas. 2004. "Staged Development in Xinjiang." *China Quarterly* 178 (June): 358–378.

Bhattacharji, Preeti. 2008. "Uighurs and China's Xinjiang Region." Council on Foreign Relations website. http://www.cfr.org/publication/16870/#1.

China. 1981. *Dang de minzu zhengce wenxian ziliao xuanbian* [Selected Documents on the CCP's Minority Policies]. Beijing: The Institute of Nationalities Studies of the Chinese Academy of Social Sciences.

———. 1987. *Sulian minzu wenti wenxian xuanbian* [Selected Soviet Documents on the National Question]. Beijing: Shehui kexue wenxian chubanshe.

———. 1997. *Jianguo yilai Zhongyao Wenjian Xuanbian* [Selected important documents since the founding of the PRC], 20 vols. Beijing: Zhongyang Wenxian Press.

———. 1998. *Zhonghua renmin gongheguo xiangxing jiaoyu fagui huibian 1990–1995* [Collection of current education laws and regulations of the PRC, 1990–1995]. Beijing: Renmin jiaoyu chubanshe.

———. 2002. *Zhongguo gongchangdang guanyu minzu wenti de jiben guangdian he zhengce* [The Chinese Communist Party's Basic Views and Policy on the National Question]. Beijing: Minzu chubanshe.

———. 2005a. *Guowuyuan shishi zhonghua renmin gongheguo minzu quyu zizhi fa ruogan guiding shiyi* [Interpretations of the State Council's Implementation Regulations for the PRC Law on Ethnic Territorial Autonomy]. Beijing: Minzu chubanshe.

———. 2005b. *Guowuyuan guanyu jichu jiaoyu gaoge yu fazhan de jueding* [The State Council's decision on the reform and development of the basic nine-year education] http://www.edu.cn/20051123/3162315.shtml.

Cohn, Bernard. 1987. "Census, Social Structure, and Objectifications in South Asia," in *An Anthropologist Among Historians*, ed. Bernard S. Cohn, 224–254. Delhi: Oxford University Press.

Connor, Walker. 1984. *The National Question in Marxist-Leninist Theory and Strategy*. Princeton, NJ: Princeton University Press.

Dreyer, J. Teufel. 1976. *China's Forty Millions*. Cambridge, MA: Harvard University Press.

———. 2006. *China's Political System: Modernization and Tradition*. New York: Pearson.

Fei, Xiaotong. (ed.) 1991. *Zhonghua minzu yanjiu xintansuo* [New Perspectives on Studies of the Chinese Nation]. Beijing: Shehui Kexue Press.

———. (ed.) 1999. *Zhonghua minzu duoyuan yiti geju* [The Pattern of Diversity in Unity of the Chinese Nation]. Revised Edition. Beijing: Zhongyang minzu daxue chubanshe.

Gladney, Dru. 2004. *Dislocating China: Reflections on Muslims, Minorities, and Other Subaltern Subjects*. London: Hurst.

Guo, Hongsheng. 1997. *Zhongguo yu qian sulian mizu wenti duibi yanjiu* [A Comparative Study of the National Question in China and the Former Soviet Union]. Beijing: Minzu chubanshe.

Hua, Xinzhi and Dongen Chen. 2002. *Sidalin yu minzu wenti* [Stalin and the National Question]. Beijing: Minzu chubanshe.

Jaladi, Rita and Seymour Martin Lipset. 1992–93. "Racial and Ethnic Conflicts: A Global Perspective." *Political Science Quarterly* 107 (4): 585–606.

Ledovskii, A. M. 1999/2001. *Stalin and China* [Sidalin yu Zhongguo], trans. Chunhua Chen and Zunkuang Liu. Beijing: Xinhua chubanshe.

Lenin, Vladimir Il'ich. 1967. *V. I. Lenin Selected Works*, vol. 3. New York: International Publishers.

Li, Dezhu (ed.) 1999. *Zhongyang disan dai lingdao yu shaoshu minzu* [The Third Generation of CCP Leaders and the Minority Nationalities]. Beijing: Zhongyang minzu daxue chubanshe.

Liu, Chun. 1996. *Liu Chun minzu wenti wenji* [Selected works of Liu Chun on the national question]. Beijing: Minzu chubanshe.

Liu, Lydia. 1995. *Translational Practice: Literature, National Cultures, and Translated Modernity—China, 1900–1937*. Stanford: Stanford University Press.

Luo, Guangwu. 2001. *Xin zhongguo minzu gongzuo dashi gailan* [An Introduction to Major Events in New China's Nationalities Affairs]. Beijing: Huawen chubanshe.

Mackerras, Colin. 1994. *China's Minorities: Integration and Modernization in the Twentieth Century.* Hong Kong: Oxford University Press.

Mao, Gongning. 2001. *Minzu wenti lunji* [A Collection of Articles on the National Question]. Beijing: Minzu chubanshe.

Mao, Gongning and Liu Wanqing. 2004. *Minzu zhengce yanjiu wencong* [Studies of Minorities Policy], vol. 1. Beijing: Minzu chubanshe.

Sautman, Barry. 1998, "Preferential Policies for Ethnic Minorities in China: The Case of Xinjiang," in *Nationalism and Ethnoregional Identities in China,* ed. William Safran, 86–118. London; Portland OR: Frank Cass.

Stalin, Joseph. 1975. *Marxism and the National-Colonial Question.* San Francisco, CA: Proletarian Publishers.

Strong, Anna Louise. 1937. *The New Soviet Constitution: A Study in Socialist Democracy.* New York: Henry Holt and Company.

Theodore de Bary, WM. and Irene Bloom. 1989. *Sources of Chinese Tradition: From Earliest Times to 1600.* Second edn. New York: Columbia University Press.

Tiemuer, and Wangqing Liu (ed.) 2002. *Minzu Zhengce Yanjiu Wencong, Diyi Ji* [*Studies of Minorities Policy*] Volume 1. Beijing: Minzu chubanshe.

Wang, Geliu and Jianyue Chen. 2001. *Minzu quyu zizhidu de fazhuan: minzu quyu zizhifa xiugai wenti yanjiu* [The Development of Autonomy for Minority Regions: Studies on the Issues in the Revision of Laws on Autonomy for Minority Regions]. Beijing: Minzu chubanshe.

Wang, Jianmin, Haiyang Zhang, and Hongbao Hu. 1998. *Zhongguo minzuxue shi, xia juan, 1950–1997* [The History of Ethnology in China, Part II, 1950–1997]. Kunming: Yunnan jiaoyu chubanshe.

Wen, Jin, Gongning Mao, and Tiezhi Wang. 2001. *Tuanjie jinbu de weida qizhi—zhongguo gongchan tang 80 nian minzu gongzuo lishi huigu* [The Great Banner of Unity and Progress: A Review of the CCP's 80 Years of Minority Work]. Beijing: Minzu chubanshe.

Zhou, Minglang. 2003. *Multilingualism in China: The Politics of Writing Reforms for Minority Languages 1949–2002.* Berlin/New York: Mouton de Gruyter.

———. 2006. "Globalization and language education in America and China: Bi/multilingualism as an ideology and a linguistic order," invited speech at the GSE colloquium at the University of Pennsylvania, November 30, 2006.

———. Forthcoming. "The Fate of the Soviet Model of Multinational State-Building in China," in *China Learns from the Soviet Union, 1949–present,* ed. Tom Bernstein and Hua-yu Li. Lanham, MD: Rowman & Littlefield Publishers.

Part I

Debating China's Positive Policies: Historical Antecedents and Contemporary Practice

Chapter 1

Mandarins, Marxists, and Minorities

Walker Connor

Those charged with designing or implementing policies to placate minorities ensconced in their own homelands face two levels of problems. The first, the urge of minorities to control their own destinies and therefore to resent and resist rule by outsiders, is common to all multi-homeland states. The second set of problems evolves from each country's unique experience. What follows is a discussion of both of these frameworks, within which current Chinese minority policies necessarily operate and that are apt to influence the relative success of those policies.

All but a handful of today's states are multinational. Their policies toward national minorities vary enormously. In very few cases, governments have permitted—in still rarer cases even encouraged—a homeland-dwelling people to secede. However, determination to maintain the territorial integrity of the state customarily causes governments to view secession as anathema. More commonly, if a government desires to rid the country of a minority, it pursues a policy of what is currently called "ethnic cleansing," usually through genocide, expulsion, and population transfers—employed separately or in combination. Far more commonly, however, governments accept the inhabitants within their borders as a given and introduce assimilationist programs. Such programs vary considerably in scope, complexity, degree of persuasion or coercion, timetable, and fervor of the implementers. But programmed assimilation does not have an impressive record, as we are reminded by the history of the Soviet Union. After seventy years of comprehensive and sophisticated governmental efforts to solve what was officially termed "the national question," national consciousness and resentment grew among the Soviet Union's non-Russian peoples. As a result of such failures, there has been a recent increase in the number of states that have decided to eschew assimilation as a policy in favor of decentralization, that is, the granting of cultural and/or political autonomy to national minorities. Cases in point include Belgium, Canada, Italy, Spain, and the United Kingdom. But recent developments within Switzerland—which handily represents the longest running experiment in investing minorities with meaningful autonomy—as well as developments within Belgium and Catalonia, suggest that even marked decentralization may not totally quench separatist sentiments.

In sum, although some states (such as Switzerland) have been far more successful than others (such as Sri Lanka), no multinational state can lay claim to having found the formula or formulas for peacefully accommodating ethnic heterogeneity. This volume focuses on China's policies toward its minorities. But whatever the particulars of those policies, the record of failures on the part of other states suggests that those policies face formidable barriers to achieving success. Two of the most fundamental of these barriers can be treated under the headings (1) the nature of ethnic nationalism and (2) homeland psychology.

The Nature of Ethnic Nationalism

Ethnic nationalism connotes identity with and loyalty to a nation in the sense of a human grouping predicated upon a claim of common ancestry. Seldom will the claim find support in scientific evidence. The still limited number of DNA analyses available document that the degree of ancestral purity varies substantially among peoples. Studies of the patrilineally bequeathed Y chromosome attest that nations tend to be neither genetically homogeneous nor hermetical, and analyses of the matrilineally bequeathed mitochrondrial DNA customarily attest to still greater heterogeneity and transnational genetic sharing. However, the popularly held conviction that one's nation is ethnically pure and distinct is intuitive rather than rational in its wellsprings and, as such, often defies scientific and historic evidence to the contrary.

Ethnic nationalism is often contrasted with a so-called civic nationalism, by which is meant identity with and loyalty to the state. (Until quite recently the latter was conventionally referred to as patriotism.) The practice of referring to civic consciousness and civic loyalty as a form of nationalism has spawned great confusion in the literature. Rather than representing variations of the same phenomenon, ethnic and civic loyalties are of two different orders of things. While in the case of a people clearly dominant within a state, such as the ethnically Turkish, Castilian, or Han Chinese peoples, the two loyalties may reinforce one another; in the case of ethnonational minorities, such as the Kurds of Turkey, the Basques of Spain, or the Mongols, Tibetans, and Uygurs of China, the two identities may clash. World political history since the Napoleonic Wars has been increasingly a tale of tension between the two loyalties, each possessing its own irrefragable and exclusive claim to political legitimacy.

The concept of political legitimacy inherent in ethnonationalism rests upon the tendency of people living within their homeland to resent and resist rule by those perceived as aliens. Evolutionary biologists classify xenophobia as a universal that has been detected on the part of all societies studied thus far (Brown 1991). Buttressing this finding of universality are the histories of multiethnic empires—ancient and modern—that are sprinkled with ethnically inspired insurrections. The modern state-system has proven even more vulnerable. In the 130-year period separating the Napoleonic Wars from the end of World War II, all but three of Europe's states had either lost extensive territory and population because of ethnonational movements or were themselves the product of such a movement. Ethnic nationalism's challenge to the multinational state continued to accelerate during the late twentieth century, culminating in the dissolution of the Soviet Union and Yugoslavia.

During the course of its development, the equating of alien rule with illegitimate rule came to be called "national self-determination," a phrase probably coined by Karl Marx and subsequently frequently employed by the First and Second Internationals. National self-determination holds that any group of people, simply because it considers itself to be a separate nation (in the pristine sense of a people who believe themselves to be ancestrally related), has an inalienable right to determine its political affiliations, including, if it so desires, the right to its own state. Lenin had a profound appreciation of the power of the self-determination urge, and he assigned appealing to it as a key stratagem for overthrowing governments. Consonant with his wishes, the Communist International insisted that all communist parties, prior to assuming power, must appeal to their country's national minorities by promising that upon taking power they would recognize the right of self-determination for all national groups, explicitly including the right to secede. For a time, the Chinese Communist Party (CCP) refused to follow this order, but, commencing in 1930 and continuing until final victory over the Nationalist Party (*Guomindang*, or GMD) in 1949, pledges to honor a right to secession were part of official policy.

In 1930, the CCP outlined what it termed "Ten Great Political Programs." Number five was the recognition of the right of minorities to secede. The following year an article in the Party's journal, *Bolshevik*, noted "the CCP...must advocate that the non-Han nationalities...be given the right of self-determination and even the right of secession...it is in the revolutionary interest of the Chinese Soviet Government to insure actual independence and freedom of the non-Han nationality states" (cited in Dryer 1977, 63). In accord with this policy, the fourteenth article of the 1931 founding constitution of the Soviet Government of China read:

> The Soviet Government of China recognizes the right of self-determination of the national minorities in China, their right to complete separation from China, and to the formation of an independent state for each national minority. All Mongolians, Tibetans, Miao, Yao, Koreans, and others living on the territory of China shall enjoy the full right to self-determination, i.e., they may either join the Union of Chinese Soviets or secede from it and form their own state as they may prefer. The Soviet regime of China will do its utmost to assist the national minorities in liberating themselves from the yoke of imperialists, the GMD militarists, *tusi*, the princes, lamas, and others, and in achieving complete freedom and autonomy. The Soviet regime must encourage the development of the national cultures and of the respective languages of these peoples. (Constitution of the Soviet Republic [1931] 1952, 223–224)

Sporadic promises of independence were made to those minority peoples encountered during the Long March. In 1935, the CCP noted in a declaration to the Mongols:

> [W]e recognize the right of the people of Inner Mongolia to decide all questions pertaining to themselves, for no one has the right to forcefully interfere with the way of life, religious observances, etc., of the Inner Mongolian people. At the same time the people of Inner Mongolia are free to build a system of their own choosing; they are at liberty to develop their own government, entirely separate. (Reproduced in Chang Chih-I 1951, 50–52)

In December of the same year, Mao signed a resolution that asserted: "By its own example and sincere slogans, the Soviet People's Republic tells the oppressed Mongolians and Moslems: Organize your own state" (Mao Zedong [1935] 1978, vol. 5–6, 22). In May of the following year Mao addressed the "Turkish Moslems" and other "minorities of the Northwest" thus:

> According to the principle of national self-determination, we advocate that the affairs of the Moslems must be completely handled by the Moslems themselves, that, in all Moslem areas, the Moslems must establish their independent and autonomous political power and handle all political,

> economic, religious, custom, ethical, educational, and other matters. (Mao Zedong [1936] 1978, 35–36)

References to CCP support for self-determination continued during and after World War II, and as late as November 1948 Liu Shaoqi, then a member of the Politburo and heir apparent to Mao, proclaimed that the CCP advocates "the voluntary separation of all nations" (Liu Shaoqi [1948] 1969, 127–128).

The CCP therefore continued to hold out the grail of national self-determination until the very eve of victory over the GMD. But with that victory, all talk of national self-determination and secession ceased. No mention of them was made in the provisional constitution of 1949. On the contrary, the constitution declared the need "to liberate all of the territory of China, and to achieve the unification of China" (Article 2, China 1950, 3); "to suppress all counter-revolutionary elements who... commit treason against the motherland" (Article 7, China 1950, 4); and it warned that "nationalism and chauvinism will be opposed" (Article 50, China 1950, 18). Within a month of formally declaring victory, a directive was cabled from the central party propaganda office of the New China News Agency to the Northwestern branch office:

> Today the question of each minority's "self-determination" should not be stressed any further. In the past, during the period of civil war, for the sake of strengthening the minorities' opposition to the Guomindang's reactionary rule, we emphasized this slogan. This was correct at the time. But today the situation has fundamentally changed... For the sake of completing our state's great purpose of unification, for the sake of opposing the conspiracy of imperialists and other running dogs to divide China's nationality unity, we should not stress this slogan in the domestic nationality question and should not allow its usage by imperialists and reactionary elements among various domestic nationalities... The Han occupy the majority population of the country; moreover, the Han today are the major force in China's revolution. Under the leadership of the Chinese Communist Party, the victory of Chinese people's democratic revolution mainly relied on the industry of the Han people.[1]

There is ample evidence that the CCP leadership never intended to honor the pledges made to the minorities. Indeed, from the outset, Mao had been simultaneously appealing to Han ethnic nationalism by, inter alia, pledging to never surrender any part of China's territory. To this set of pledges he remained true: Tibet, Xinjiang, and other wayward regions were soon retaken by the Red Army. In retrospect then, the leadership of the CCP had demonstrated a profound appreciation of the power that ethnonationalism—of both minority and Han variety—exerts. And they could take satisfaction in their having expertly employed this force to gain political supremacy.

On the negative side of the ledger, their very success with the minorities would be a constant reminder to the leadership that the loyalty of the minorities could not be trusted, that many of them harbored a desire to secede from China. A strong case could be made that this distrust of the minorities is reflected in subsequent minority policies, surfacing on occasion in undisguised form, such as during the Great Leap Forward and the Cultural Revolution. Conversely, the record of the CCP's broken promises to recognize the right of independence for minorities provides good reason for the non-Han peoples to distrust the authorities. The ruling elite continues to present itself as the vanguard of the Chinese Communist Party, the present incarnation of a CCP leadership stretching back to the 1920s.[2] As such, they have become the legatees of the party's previous minority policies.

Homelands and Homeland Psychology

In addition to being a multinational state, China is also a multihomeland state. Again, this is a characteristic shared by most states. With the principal exception of a few immigrant societies such as Argentina, Australia, and the United States, the land masses of the world are divided into ethnic homelands, territories with names reflecting a particular people. Catalonia, Croatia, England (from Engla-land: land of the Angles), Euskadi ("Basque Homeland"), Finland, Iboland, Ireland, Uyguristan (literally "land of the Uygurs"), Mongolia, Nagaland, Pakhtunistan, Poland, Scotland, Swaziland, Sweden, Tibet, and Uzbekistan constitute but a small sampling.

To the people who have lent their name to the area, the homeland is much more than a territory. The emotional attachment is reflected in such widely used descriptions as "the native land," "the fatherland," "this sacred soil," "the ancestral land," "this hallowed place," "the motherland," "land of our fathers," and, not least, "the *home* land." In the case of a homeland, territory becomes intermeshed with notions of ancestry and family. This emotional attachment to the homeland derives from perceptions of it as the cultural hearth and, very often, as the geographic cradle of the ethnonational group. The important point is that the populated world is subdivided into a series of perceived homelands to which, in each case, the indigenous ethnonational group is convinced it has a profound and proprietary claim.

As a consequence of the sense of primal ownership that an ethnonational group harbors toward its homeland, non-members of the ethnic group within the homeland are viewed as aliens ("outsiders"), even if they are compatriots. They may be endured, even treated equitably. Their stay may be multigenerational. But they remain outsiders or settlers in the eyes of the homeland people, who reserve what they deem their inalienable right to execute their primary and exclusive claim to the homeland whenever they desire and, of course, have the capability. In sum, the notion of a primal ownership holds that only the members of "our people" have a "true right" to be here.

Given that the land masses of the world are divided into some three thousand homelands over which the political borders of fewer than two hundred countries have been superimposed, it is hardly surprising that most states are not just multinational but also multihomeland. This is of the greatest significance when assessing the probable political instability of tomorrow's world, for the demands of ethnonational movements tend to be coterminous with their homeland. In terms of geography, it is for the homeland that ethnonational groups demand greater autonomy or full independence. It is over Euskadi, Corsica, Kashmir, Nagaland, and Tibet that the Basques, the Corsicans, the Kashmiris, the Nagas, and Tibetans demand greater control.

When analyzing minority policies it is essential to differentiate between homeland-dwelling peoples and minorities created by migration, that is, minorities not living within their homelands. In general, only the former tend to demand independence or autonomy. Consider, for example, the immigrant, fundamentally non-homeland United States. That country has certainly not been free of ethnic problems. But it is equal rights and opportunities, not questions of autonomy or separatism, that dominate minority/majority relations there. Questions concerning autonomy arise within the United States only in relation to its relatively few homeland peoples: Hawaiians, those Amerindian peoples who have elected to remain on "reserved" Indian lands, and those Eskimos or Inuit in settled communities within Alaska. The point is that it is the integrity of the multihomeland state that is challenged by ethnonationally inspired movements and that analogies should not be drawn between their problems and the experiences of uni-homeland or non-homeland states.

The administrative divisions of the former Soviet Union corresponded, at least roughly, with the country's major homelands. Fourteen of the fifteen socialist republics comprising the Soviet state bore the ethnonym of a national group and approximated the contours of that group's homeland. And the implosion of the Soviet state corresponded to the pattern we have ascribed to homeland-dwelling people. It was precisely over their homeland—over mother Armenia, Estonia, Georgia, Latvia, Lithuania, and so on—that the non-Russian peoples demanded of Gorbachev greater control.

Beijing has also created administrative units with official designations containing the names of a specific people (e.g., the Tibetan Autonomous Region). Unlike the Soviet Union, Beijing did not create national republics with the theoretic (though definitely not practicable) right of secession. Indeed, immediately after the CCP's assumption of power, the term nation, which had been regularly applied to the Mongols, Tibetans, Uygurs, and others in documents recognizing their right to secession, was abandoned in favor of the term nationality. The change was significant, for in the Marxist-Leninist lexicon only nations have a right of self-determination and secession; nationalities, a lower category, do not. Consonant with the new nomenclature, the Chinese therefore avoided creating any republics, creating instead a three tiered system of autonomous regions (ARs), autonomous districts or prefectures (ADs), and autonomous counties (ACs). From the outset, the borders of these units were purposefully designed to violate, not reflect, homelands. Borders were drawn so as to omit large parts of the homeland (among the ARs, particularly flagrant in the cases of the Guangxi Zhuang and Tibetan ARs) and/or to incorporate large areas outside the homeland populated by others (particularly flagrant in the case of the Guangxi Zhuang and Xinjiang Uygur ARs). The purely expediential regard in which the authorities hold these borders is suggested by the following account of the Mongolian AR:

> [T]he first autonomous region was that of Inner Mongolia, created in 1947 while the struggle for supremacy over China was still in progress. As originally delineated, the AR housed two to four times as many Han as it did Mongols. In 1955 the authorities added the predominantly Han provinces of Suiyan and Chahar (Huhehot) and part of a third, Ningxia, thereby increasing the ratio of Han over Mongols to eight to one. In 1969 the AR's territory was just as abruptly decreased by approximately one-third, as huge segments of territory were severed from both its eastern and western flanks. Erstwhile Mongolian inhabitants of their own AR found themselves part of either Heilungkiang, Kirin, Liaoning, or Kansu Province, or, more paradoxically, part of the Ningxia Hui AR. The partition not only further numerically disadvantaged those Mongols within the truncated AR, but left a majority living outside its borders. The impermanence of such borders was again underlined in 1979 when, as part of the post-Cultural Revolution rewooing of the minorities, the region was returned to its 1955–1969 size. While this alteration at least had the effect of placing a majority (about 70 percent) of all of China's Mongols within their own autonomous region, it left the Mongols but a small fraction of the region's total population. (Connor 1984, 323)

Numerous other national groups found themselves from the very beginning a minority in a so-called autonomous unit bearing their ethnonyms.[3]

The flagrant gerrymandering of homelands has therefore led to their ethnic dilution, and this evisceration has been furthered by policies mandating or encouraging the in-migration of the Han (official name given to the majority in China). The largest resettlements appear to have occurred under the "send down" (*xia fang*) campaign that was initiated in 1958 during the Great Leap and that survived both the Great Leap and the Cultural Revolution (1966–76). Following the Cultural Revolution, such mandated programs lost favor, and many Han took advantage of the changed climate to emigrate from the non-Han homelands to the economically booming, traditionally Han dominated areas to the east. The lasting impact of the earlier migrations has often been great, however. Thus, between 1949 and 2000, the percentage of Han in the population of Xinjiang increased from some 7 to 40.[4] Moreover, while eschewing direct control over migration the authorities are still able to encourage influxes into non-Han homelands (1) by making settlement in these areas more attractive through such devices as improving the infrastructure, extending affirmative action benefits to everyone in the homelands regardless of ethnic heritage, and so on and (2) by making the homelands more readily accessible. With regard to the latter, in a truly remarkable undertaking that could not be justified in terms of anticipated financial return, a railroad spanning "the roof of the world," connecting Lhasa with eastern China, was opened in 2006. According to one account: "By some estimates, the new train will carry as many as 900,000 people to Tibet each year, with the newcomers overwhelmingly consisting of members of China's Han majority, many of whom will opt to stay, further dampening demands for independence and diluting Tibet's spiritual culture" (French 2005).

The in-migration of ever greater numbers of Han may indeed dampen the demands but certainly not the desire for independence. The experiences of a large number of other societies document that the sense of primal ownership of a homeland causes a surge in immigration to be followed by a rise in xenophobic and separatist sentiment.[5] On the one hand, it is evident that significant dilution of the homeland people lessens the risk of the homeland developing into an effective base for antistate activities. The Mongols, for example, are only a small fraction of the population of the Inner Mongolian Autonomous Region, making it less likely that they can develop an effective, indigenous separatist movement than would be the case if they were a majority. On the other hand, in-migration raises the level of general resentment felt toward the Han, as well as of general distrust of Beijing's policies.

The CCP and Affirmative Action Policies

Lenin had designed a blueprint for minority policy in a postrevolution situation. The long-term goal was assimilation or, to use his terminology, the fusion, merging, or coming together of nations. Similarly, Beijing usually avoids using the word for assimilation, preferring *ronghe*, which connotes a melding together or an amalgamation. The desire to avoid the term assimilation rested on the twin assertions that (1) assimilation is better suited to capitalist societies, wherein the relations among national groups are characterized by inequality and oppression and where coercion is the principal means for bringing about acculturation; and (2) assimilation usually refers to absorption by the state's dominant national group and is therefore a euphemism for Russification, Sinification, Vietification, Serbification, and the like. Marxist-Leninists differentiate their approach to the national question by claiming, first, that national relations within a Marxist society are predicated upon absolute national equality and, second, that the process of blending together is fully voluntary, devoid of any element of coercion.

Campaigns to bring about assimilation have been plentiful throughout history, but what was unique to Lenin's scheme was his strategy to achieve it through the seemingly contrary policy of pandering to the more overt cultural distinctions of minorities. This dialectical approach to achieving assimilation flowed from his perception of nationalism as the outgrowth of past discrimination and oppression. The resulting suspicion and mistrust held by the minorities was to be overcome by a period of cultural pluralism during which the more obvious manifestations of each nation's uniqueness, most especially its language, were to be nurtured by the state.

In time this policy of promoting pluralism came to be known as "the flourishing of the nations." Lenin reasoned that as the policy of cultural pluralism dissipated the antagonisms and mistrust that had previously estranged nations, those human units would naturally move closer together, a process that became known in the official Marxist lexicon as "the rapprochement" or "coming together" of nations. The process would continue until a complete blending was achieved, and a single identity had emerged.

In Lenin's scheme, then, the period of the flourishing of nations would foster the process of *e pluribus unum*. To Lenin, language and other overt manifestations of national uniqueness were construed, on balance, as useful conveyers of the messages emanating from the party. In and by themselves they were merely forms. It was the party, acting through the state, that would give them content. Forms did, nevertheless, have an important role to play in enhancing the receptivity accorded to the messages. Lenin and his successors believed that the minorities would not see the party's programs as alien, identified with the state's dominant ethnic element, if they came dressed in the local tongue and other appropriate national attire. Employing the individualized national forms would convince the people that the Party's minority policies were not just a new guise for assimilation by the dominant group. In 1925 Stalin would confer upon this entire approach to the national question the official abbreviated title of "national in form, socialist in content."

Although adopted by Yugoslavia, the Soviet Union, and other Marxist states under Soviet sway, Lenin's plan for achieving assimilation via the dialectical route of proceeding through a lengthy period of national flourishing was nowhere given a full testing. Anxious to accelerate the process of assimilation and worried that Lenin could be wrong—that permitting and encouraging cultural autonomy might in fact encourage minority nationalism and separatist sentiment—governments have vacillated. They have also been anxious to accelerate the process of assimilation. Periods of the flourishing of nations have alternated with periods of promoting rapid acculturation and assimilation. China proved no exception, sometimes leaning in one direction, sometimes the other. Code words and signs guide the outsider in following the fluctuations. In periods characterized by greater tolerance of national peculiarities, Great Nation (i.e., Han) chauvinism is described as the greatest threat; when more rapid acculturation/assimilation is the current goal, local nationalism becomes the bête noir. The acclaim or invective heaped upon legendary heroes of the minorities—figures such as the Mongol Chinggis Khan—is also a key. While the Khan's momentary standing with the Beijing authorities might seem a somewhat frivolous, minor issue, that standing has in fact been one of the most reliable and easily read barometers of official policy on the national question.

The early approach of the CCP to the country's national question showed great respect for national peculiarities. Although the 1949 constitution noted that both "[n]ationalism and chauvinism shall be opposed," its Article 53 made clear that the flourishing of the nations was to be condoned: "All national minorities shall have freedom to develop their spoken and written languages, to preserve or reform their traditions, customs, and religious beliefs. The people's government shall assist the masses of all national minorities in their political, economic, cultural, and educational development" (China 1950, 19). This was the period of the Chinese version of "indigenization," when the strategic thinking of the leadership on the national question was colorfully captured in the official description of its national policy as "No Struggle." Not only were local languages and other aspects of culture promoted, but differential time schedules for achieving socialism on the part of the Han and non-Han peoples were officially approved. In practice the latter meant that minorities could be at least temporarily excused from unpopular "reforms" such as land collectivization. During the period, Chinggis Khan was lionized and much publicity was given to the return of his bones to the mausoleum from which they had been removed years earlier to protect them from Japanese invaders. Their return was heralded as illustrating "the profound concern of the Chinese Communist Party and of Chairman Mao for the minority nationalities" (cited in Dryer 1972, 178).

This period of courtship lasted from 1949 to 1957, but subtle shifts within the period are detectable. At first glance, an official summary of the experience of the first four years of the national policy appears to swing further in the direction of the flourishing of nations than had the 1949 constitution, in that it proclaimed that Great Hanism "at the moment constitutes the major danger for the proper relationships among various nationalities" ("Basic Summarization of Experiences..." 1953, 16). However, the overall thrust of the document was in fact in the other direction. It prescribed "active assistance" for those who desired to learn Han rather than their own language, and cautioned those who believed that autonomy would lead to the elimination of Han cadres and settlers in minority regions (p. 17). While noting the tactical advantage of carrying out the party's tasks through "the use of appropriate national forms" (p. 19), it unambiguously warned that "respect for national forms is not to be carried to the stage of the preservation of even such forms which obstruct the progress and development of the nationality" (p. 20).

The 1953 report can therefore be viewed as a decision to move slightly away from the erstwhile emphasis upon national flourishing. Flourishing would be continued but in a more carefully circumscribed manner; assimilation was to be encouraged but not pushed. This slight leaning in favor of assimilation continued over the next two years. The preamble of the 1954 state constitution promised that the government would "pay full attention to the special characteristics in the development of each nationality" (China 1962, 64). On the other hand, the document did not single out Han chauvinism as the principal threat to an effective national policy as had the report issued the previous year. The 1954 constitution pledged instead "opposition to both big-nation chauvinism and local nationalism" (p. 64).

An abrupt shift back toward the flourishing stance of the 1949–53 period was signaled by Mao himself in April 1956. In an address to the Politburo, Mao listed ten major problem areas (that he identified as "the ten great relationships"), among which was numbered "the relationship between the Han nationality and the national minorities" (Mao Zedong [1956] 1974, 74). Though Mao maintained that on this issue "our policy is stable," it is evident that he intended a relaxation of controls over the minorities (p. 74). Unlike the constitution, which was itself less than two years old, Mao's speech singled out Great Han chauvinism as the principal adversary: "Our emphasis lies on opposing Han chauvinism. Local nationalism exists, but this is not the crucial problem. The crucial problem is opposition to Han chauvinism" (p. 74).

Just as abruptly, however, the pendulum swung all the way back toward the assimilationist pole late in the year following Mao's famous Hundred Flowers Bloom speech, a speech urging what he assumed would be minor criticisms of the system. But the speech evoked an unanticipated outburst of discontent on the part of minorities, such as: "The Chinese Communist Party's policy of regional autonomy for the nationalities is that of 'divide and rule'" (*Yunnan Daily* 1957). The present system of regional autonomy is "as useful as a deaf ear" (cited by Dryer 1972, 224). "All the principal responsible persons of Party committees at various levels are Han Nationals [which is] contradictory to having autonomous nationalities run their own affairs" (*People's Daily* 1957). Cadres of the local ethnic groups are "traitors to their nationality" and "jackals serving the Han" (cited in Dryer 1972, 224). "The minority nationalities run the house but the Han people give the orders" (*Guangming Daily* 1958). "Many rights in theory, few in practice" (cited in Dryer 1972, 224).

The publication of such complaints heralded a radical swing in Chinese national policy. A full-scale attack on local nationalism was launched. The program blended well with the policies of the Great Leap Forward (1957–58), which was inaugurated shortly thereafter. As applied to the national question, the Great Leap can be thought of as an attempt to complete in one all-out, intensive effort the process described in the Communist Manifesto as "national differences and antagonisms between peoples are daily more and more vanishing." In the words of one Chinese authority, "the less *[sic]* differences among peoples, the faster development can be" (p. 241). Struggle was to be waged against all symptoms of non-Han national individuality. Different tempos on the road to socialist construction were no longer to be tolerated. National dress, dance, and the like were discouraged. The Chinese language was introduced as the language of instruction in all grades. Intermarriage with Han was encouraged and, in some areas, reportedly forced. Huge numbers of Han were ordered into minority areas. Multinational communes were created that ensured that the minorities could not remain aloof from acculturating and assimilating influences. The study of national culture and national histories was ordered curtailed. Chinggis Khan was now reviled as a brutal despot. Regional autonomy was attacked as "outmoded" and "unnecessary."

These extreme measures provoked extreme reactions from the minorities, soon leading to their moderation. There followed a period of rivalry between extremists (those identified with Mao who were pushing for rapid assimilation) and the gradualists (identified with Liu Shaoqi, Zhou Enlai, Zhu De, and Deng Xiaoping). Policy between 1958 and 1965 reflected the gradualist approach. The gradualists decentralized the multinational communes and re-extended recognition of the unique traditions of national groups in matters of eating habits and so forth. The drive to have *Putonghua* (Mandarin) introduced everywhere as the language of instruction was quietly dropped. Traditional leaders who had been scheduled to be purged under the Party line that "the national question is a class question" were now promised "a bright future" by the country's leading newspaper; cadres were instructed to curb their zeal (Hyer and Heaton 1968, 24).

The rapid assimilationist faction continued to vie for power, however. With the inception of the Cultural Revolution in 1966, it was evident that they had gained the upper hand. Regional autonomy, toleration of national differences, different tempos for achieving socialism, and cooperation with local leaders were all now violently attacked, and demands for proceeding with immediate assimilation were again raised. Unlike the period of the Great Leap, those favoring rapid assimilation could now savor the humiliation and purging of the most renowned figures who had publicly defended a gradualist approach to the national question. The list included no less a personage than the chief-of-state Liu Shaoqi. The most prominent gradualists were not, of course, purged solely or even principally because of their attitude toward the national question. It was their reluctance to prescribe radical therapy for all aspects of Chinese society that enraged their opponents. However, given the relative numerical insignificance of the non-Han peoples, it is remarkable how much emphasis was placed upon the alleged deviations of the gradualists with regard to the national question. Thus, the general secretary of the party, Deng Xiaoping, was accused of abetting the 1959 Tibetan revolution. And a pamphlet aimed at bringing down the chief-of-state was entitled *Completely Purge Liu Shaoqi for His Counter-Revolutionary Revisionist Crimes in United Front, Nationalities, and Religious Work* (1967). After citing from Liu's writings of 1937, 1948, and 1954—all of which had reflected what at the time was the current position of Mao and the party—the authors charged him with being a proponent of "national separatism."

Those who were purged for favoring a more gradual approach to the national question included a number of non-Han, some of whom had long been among the most honored and trusted members of the CCP elite. Typical of the allegations were charges that they had promoted "national splittism" and had been "trying to establish an independent kingdom."[6]

Those favoring rapid assimilation remained preeminent from 1966 until 1971. As the latter year wound down, a number of public disclosures and announcements suggested that yet another shift in the power struggle had occurred. Clues included a series of personnel changes in the leadership of the autonomous regions. A *People's Daily* article called for more minority cadres, and included numerous uncritical references to the cultural differences among national groups. During 1972 indications of a changing national policy multiplied. The Central Nationalities Institute, closed down during the Cultural Revolution, was reopened; national forms were once more described in favorable terms; and selective quotations by Mao from the pre-Hundred Flowers period when he had supported the flourishing of nations began to slip into print. Even more symptomatic of a shift was the re-elevation to good standing of Deng Xiaoping, along with a number of non-Han leaders such as Ulanhu who had been purged during the Cultural Revolution.

An article that appeared in the *Bejing Review* in mid-1974 highlighted how far the pendulum had swung away from rapid assimilation: it suggested that the national peculiarities of the minorities would outlast even the eradication of classes and the worldwide victory of communism—that is, they would survive indefinitely.

> From the perspective of long-term historical development, the integration of nations and the extinction of nations conform to the law of historical development. But Marxist-Leninists maintain that the elimination of classes will come first, followed by the elimination of the state and finally that of nations. Lenin pointed out that mankind "can arrive at the inevitable integration of nations only through a transition period of the complete emancipation of all oppressed nations." Referring to Lenin's attitude towards the problem of nationalities, the great Marxist-Leninist Stalin pointed out that "Lenin never said that national differences must disappear and that national language must merge into one common language within the borders of a single state before the victory of socialism on a world scale. On the contrary, Lenin said something that was the opposite of this, namely, that 'national and state differences among peoples and countries . . . will continue to exist for a very, very long time even after the dictatorship of the proletariat has been established on a world scale.'" (*Beijing Review* 1974, 18)

The situation of the minorities during this period—a period that continued until 1980 and transcended the death of Mao in 1976—was at least as advantageous as, and probably surpassed, that of the period immediately following the CCP's assumption of power. At the opening of the Fifth National Congress, convened in early 1978 to approve the new constitution, Premier Hua Guofeng reestablished "great nationality Chauvinism" as the principal enemy; affirmed that "regional national autonomy must be conscientiously implemented"; and pledged to "try very hard to train cadres from minority nationalities," to guarantee "without fail" the rights of minority peoples to equality and autonomy, and "to stress the use and development of the spoken and written languages of the minority nationalities" (1978 constitution reprinted in *Beijing Review* 1978, 8–40).

The new constitution underlined the renewed significance ascribed to the minorities. The preamble alone contained two references to "all our nationalities" and another to "all the nationalities." Article 4 confirmed the equality of all nations and the right of regional autonomy for compact communities. Unlike the 1975 constitution, the new document also conferred upon the minorities not just the freedom to use their own languages but also the freedom "to use and develop their own spoken and written languages." It further signaled an official return to the flourishing of nations by granting to minorities the freedom "to preserve or reform their own customs and ways." In addition, Article 40 pledged "the higher organs of the state [to] take into full consideration the characteristics and needs of the various minority nationalities, [and to] make a major effort to train cadres of the minority nationalities."

Adoption of the constitution was followed by additional evidence that national flourishing was to be encouraged and assimilation to be played down. In October 1979 the Nationalities Committee of the National People's Congress, abolished during the Cultural Revolution, reconvened. In an address to the committee, the rehabilitated Mongolian leader Ulanhu pointed out that "the socialist stage is a time in which all nationalities develop and flourish." He also made a strong case for meaningful autonomy, noting that "autonomous organs should not exist in name only" (*Xinhua News Agency* 1979). At a more practical level, the government reopened mosques, and for the first time offered university examinations in the local languages, though with the proviso that minority students study intensive *Putonghua* for the first year. A number of non-Han were appointed to high, if largely showcase, positions, and in May 1980, Chinggis Khan was once again rehabilitated when *People's Daily* praised him as an "outstanding military strategist and statesman."

By this time however, the leadership had decided to rein in those in favor of an indefinite postponement of assimilation. In early 1980, *Red Flag* and all major newspapers carried a speech by the deceased Zhou Enlai, characterized as an "article, which expounds the national policy of the Chinese Communist Party, [and] is a Marxist work of immediate significance" (Zhou Enlai [1957] 1980, No. 9, 14). The speech reportedly delivered on August 4, 1957, was said to have been refused publication in 1958 (the period of the Great Leap Forward) and had subsequently been "suppressed for over 20 years." On the one hand, the wording of Zhou's testament on the national question assured the minorities that there was no intention of reintroducing the extreme assimilationist policies of the Great

Leap and Cultural Revolution: Han chauvinism was depicted as a more pressing danger than local nationalism, and both were described as "contradictions among the people" rather than between the people and enemies of the people. It upheld achieving socialism through differential tempos. Some nationalities were to be spared birth control requirements applied to the Han, minority languages were to be encouraged, and peoples without a written language were to be assisted in developing one. It also said that the language of the dominant group in an autonomous area was to become "the area's first language." Furthermore, indigenous cadres should represent "a proper ratio of the cadres" in autonomous areas, and "the customs and habits of all nationalities must be respected." In sum, national flourishing was reaffirmed. On the other hand, however, while approving of regional autonomy, Zhou's piece had also defended the past gerrymandering and resettlement policies that had left all autonomous units severely diluted from an ethnonational aspect. Moreover, the article had emphasized that Han cadres were to be important fixtures in all autonomous units, particularly in positions calling for what he termed "leading cadres." And most significantly, Zhou had openly championed the progressive nature of assimilation in the absence of coercion:

> The Han are so numerous simply because they have assimilated other nationalities...Assimilation is a reactionary thing if it means one nation destroying another by force. It is a progressive act if it means natural merger of nations advancing toward prosperity. Assimilation as such has the significance of promoting progress...The Hui are so huge in number just because they have succeeded in absorbing people from other nationalities. To absorb and expand—what's wrong with that? (pp. 19–20)

In effect, then, Zhou's words heralded a renunciation of both those desiring rapid assimilation and those desiring to continue minority cultural distinctiveness indefinitely. The new "plague-on-both-their-houses" position was reflected in the 1982 constitution, which is still in force. The preamble noted that "it is necessary to combat big-nation chauvinism, mainly Han chauvinism, and also necessary to combat local-national chauvinism" (*Constitution of the People's Republic of China* 1982). Moreover, while the new constitution did reserve to the indigenous peoples a number of showcase positions in the government of the autonomous units, it also introduced for the first time in a constitution a harsh warning against any separatist activities: Article 4 read in part that "any acts that undermine the unity of the nationalities or instigate their secession are prohibited." As of 2006, all of these sections had escaped the amendment process and were therefore still official policy.

Pity the Policy Makers

China's designers and implementers of national policy are thus left with, at best, nebulous guidelines: promote assimilation but avoid overly irritating national sensibilities. In practice this has meant continuing, inter alia, the transparently gerrymandered autonomous divisions, ostensibly continuing to offer the minorities a right to education in their mother tongues while curtailing the availability of textbooks and instruction in the language, most notably, at the tertiary level (Bilik 1998; Dwyer 1998), and placing members of minorities in positions of high visibility while denying them positions of real power. In 1998, for example, all twenty-two members of the Politburo were Han (Sautman 1998, 94–95). Earlier attempts to employ devices to create the illusion of autonomy while pursuing assimilation failed, as demonstrated by the harvest of opinion generated by the Hundred Flowers Program. And unhappiness on the part of the minorities since 1980—most glaringly but far from exclusively among the Tibetans and Uygurs—suggests this most recent experiment is also failing to co-opt the principal non-Han people.

From the perspective of those charged with designing and implementing policy, it would be far less onerous to successfully carry out their responsibilities if the controlling policy was either rapid assimilation or meaningful rather than illusory autonomy. Assimilation can be pursued through such heinous devices as broadly dispersing peoples outside their homelands, enforced intermarriage, and the like. Switzerland, Finland (the Aland Islands), and a few other current states suggest how policies of meaningful autonomy might accommodate ethnonational heterogeneity. But pursuing assimilation while coaxing voluntary cooperation from a minority would test the ingenuity of an oracle. Promoting a noncoercive policy of assimilation within China is particularly problematic because of the deep distrust of the authorities, the result of a history of broken promises to honor the self-determination of the minorities, including the right to secede, and of erratic policy fluctuations, when earlier programs encouraging the flourishing of nations were abruptly terminated in favor of assimilation.

Notes

1. As cited in Dittmer and Kim 1993, 275. The directive markedly undervalues the contribution of minorities to the survival and success of the CCP. The route of the Long March (1934–35) traversed the territory of minorities, and CCP's promises to support the independence of the peoples encountered was a

key in ensuring their relative indifference to the CCP's passage and sometimes in gaining their active support for the CCP's cause. From 1927 until the evacuation of the Yan'an headquarters twenty years later, Mao was operating in or near minority areas and the requirements of successful guerilla warfare made the neutrality and support of the minorities seemingly indispensable during most of this period. For details, see Walker Connor (1984).

2. During 2005–2006, the CCP leadership conducted an eighteen-month reindoctrination movement requiring its seventy million members to restudy speeches by Mao and Deng Xiaoping, as well as the Party's constitution. Interestingly, however, officially approved new history books released in 2006, in sharp contrast with previously issued histories, played down the role of Mao and the Party in the evolution of Chinese history.
3. At the level of autonomous divisions, several of the official designations included the names of two or more ethnic peoples, thereby reducing the risk of the AD becoming a focus of ethnonational emotion.
4. See the data in table 10.1 in chapter ten by Rong Ma in this volume. The Chinese leadership had thus bested Stalin's 1949 advice that the CCP increase the percentage of Han in Xinjiang to 30. Also see Minglang Zhou's chapter two.
5. For numerous examples, see Connor (2001), 64–68.
6. The pamphlet, dated April 1967, was attributed to the "Red Army" Corps of the Kangda Commune of the Central Institute for Nationalities.

References

"Basic Summarization of Experiences in the Promotion of Autonomy in National Minority Areas." 1953. *Current Background*, No. 264. Hong Kong: U.S. Consulate General, October 5, 1953.

Beijing Review. 1974. No. 17 (July 19): 18.

Bilik, Naran. 1998. "Language Education, Intellectuals and Symbolic Representation: Being an Urban Mongolian in a New Configuration of Social Evolution." *Nationalism and Ethnic Politics* 4 (Spring/Summer): 47–67.

Brown, Donald. 1991. *Human Universals*. Philadelphia: Temple University Press.

Chang, Chih-I. 1951. *The Party and the National Question in China*. Trans. George Moseley. Cambridge, MA: MIT Press.

China. 1950. *The Common Program and Other Documents of the First Plenary Session of the Chinese People's Political Consultative Conference*. Beijing: Foreign Language Press.

———. 1962. *Liu Shao-chi: Report on the Draft Constitution of the People's Republic of China/Constitution of the People's Republic of China*. Beijing: Foreign Language Press.

Completely Purge Liu Shaoqi for His Counter-Revolutionary Revisionist Crimes in United Front, Nationalities, and Religious Work. 1967. Beijing: Red Army Corps, Kanda Commune of the Central Institute for Nationalities.

Connor, Walker. 1984. *The National Question in Marxist-Leninist Theory and Strategy*. Princeton: Princeton University Press.

———. 2001. "Homelands in a World of States," in *Understanding Nationalism*, eds. Montserrat Guibernau and John Hutchinson, 64–68. Oxford: Blackwell.

"Constitution of the Soviet Republic, November 7, 1931." 1951. in *A Documentary History of Chinese Communism*, eds. Conrad Brandt et al., 220–224. London: Allen and Unwin.

Constitution of the People's Republic of China. 1982. http://www.hkhrm.org.hk/english/law/const02.html.

Dittmer, Lowell and Samuel S. Kim. 1993. "Whither China's Quest for National Identity?" in *China's Quest for National Identity*, eds. Lowell Dittmer and Samuel S. Kim, 237–290. Ithaca: Cornell University Press.

Dryer, June. 1972. "Chinese Communist Policy Toward Indigenous Minority Nationalities." PhD dissertation, Harvard University.

———. 1977. *China's Forty Millions*. Cambridge: Harvard University Press.

Dwyer, Arienne. 1998. "The Texture of Tongues: Languages and Power in China." *Nationalism and Ethnic Politics* 4 (Spring/Summer): 68–85.

French, Howard. 2005. "Railway to Tibet Is a Marvel But China is Mum." *New York Times*, September 9, 2005. http://www.nytimes.com/2005/09/09/international/asia/09golmud.html?scp=9&sq=Howard%20French%20Tibet&st=cse#.

Guangming Daily. 1958. January 17, 1958. Beijing.

Hyer, Paul and William Heaton. 1968. "The Cultural Revolution in Inner Mongolia." *China Quarterly* 36 (October–December): 124.

Liu Shaoqi. 1948. "Internationalism and Nationalism," in *Collected Works of Liu Shaoqi, 1945–1957*, 127–128. Hong Kong: Union Research Institute.

Mao Zedong. 1935. "Resolution on Current Political Situation and Party Tasks, 5/25/36," In *Collected Works, 1917–1949*, vols 5–6, 22. Hong Kong: U.S. Joint Publications Research Service (JPRS 71911), 1978.

———. 1936. "Declaration of the Soviet Central Government to the Moslem People, 12/25/35," in *Collected Works, 1917–1949*, vols 5–6, 35–36.

———. 1956. "On the Ten Great Relationships." -iIn *Chairman Mao Talks to the People: Talks and Letters, 1956–1971*, trans. and ed. Stuart Schram, 61–83. New York: Pantheon Books, 1974.

People's Daily. 1957. December 6, 1957. Beijing.

Sautman, Barry. 1998, "Preferential Policies for Ethnic Minorities in China: The Case of Xinjiang," in *Nationalism and Ethnoregional Identities in China*, ed. William Safran, 86–118. London; Portland OR: Frank Cass.

Xinhua News Agency. 1979. October 17, 1979. Reprinted in Foreign Broadcast Information Service, October 19, 1979.

Yunnan Daily. 1957. August 25, 1957. Kunming.

Zhou Enlai. 1957. "Some Questions on Policy Toward Nationalities." Reprinted in *Beijing Review* No. 9 (March 3, 1980): 14–23 and No. 10 (March 10, 1980): 18–23.

Chapter 2

Tracking the Historical Development of China's Positive and Preferential Policies for Minority Education: Continuities and Discontinuities

Minglang Zhou

China's policies on equal rights for minority education are part of a broad spectrum of rights for minorities guaranteed in the constitutions of the People's Republic of China's (PRC), beginning with its provisional constitution of 1949 and included in subsequent constitutions up to the present (1954, 1975, 1982). Questions remain, however, about how and to what extent these constitutional rights are actually put into practice (Zhou 2004). These constitutional guarantees cover political rights (e.g., autonomy, proportional representation in government and legislatures), linguistic rights, and rights to education, among others (see Article 4 and 112–120 of the PRC Constitution; for discussions, see Dreyer 2006, 167–176; Mackerras 1994, 145–166; Zhou 2003, 43–88; for Marxist and Leninist concepts, see Connor 1984, 208–239 and chapter one of this volume). While "affirmative action" or "preferential policies" are the terms most familiar to Americans when they talk of equal rights for minorities, here I use "positive (action) policies" instead. Positive polices in the international context typically refer to a much wider range of minority rights, such as those in the PRC Constitution, than the narrower limits of affirmative action in the United States. Positive policies are adopted to make up for inequalities between minorities and the majority—inequalities for which the majority are historically and morally responsible. Given this definition, in this chapter I use the term positive policies when they are truly positive but keep preferential policies for those that are not aimed at achieving justice and equality in redress of perceived historical or moral failures.

Western scholarship on the PRC's policies for minority education and the question of equal rights has developed in two phases. First, Western scholars thought that the Soviet model of positive policies for minorities was the general blueprint for the PRC's policies (Connor 1984, 87–88; Dreyer 1976, 43–60; 2006, 283–285). The Soviet constitution extended political rights and equal rights to all nationalities (see Articles 35 and 123 of the USSR 1936 Constitution; Ogden and Perelman 1960), and the language on the rights was, indeed, largely copied by the Chinese Communist Party (CCP) early on in the 1930s (see China 1981, 20; Connor 1984, 73–74). During the second phase, Western scholarship began to concentrate specifically on equal rights for China's minorities in education and in other areas as well (see Postiglione 1999; Sautman 1998). Now, scholars suggest a more direct link between policies in the Soviet Union and the PRC, and the latter's policies are considered a variant of the former's (Sautman 1999). Scholars have also speculated that the PRC copied both Soviet theory and practice, in light of the fact that some Soviet ethnologists were involved in China's work on minority issues in the 1950s (Dreyer 2006, 284). However, none of the earlier mentioned studies demonstrates clearly to what extent the PRC government adopted its positive policies from the Soviet Union and to what extent the PRC government carried them over from imperial China and Republican China (1912–49).

In this chapter, I try to bring clarity to these questions of historical legacies by identifying the continuities and discontinuities in China's tradition of making and implementing positive and preferential policies for minority education. I first examine policies and practices during imperial and Republican China, and then show exactly what the PRC has copied from the Soviet Union. I argue that (1) the PRC government has continued imperial and Republican China's practices, but has abandoned the traditional need-based principle, and that (2) the PRC government has replaced traditional China's need-based principle with the Soviet Union's principle of equal rights for all minority groups. Thus, following the Soviet Union, the CCP and the PRC government have taken a group approach to equal rights for minority education, in the process constraining the flexibility and capacity to address individual cases, and, ironically, creating new configurations of inequality, such as among different nationalities and between urban and rural residents within the same minority group.

China's Tradition of Positive and Preferential Policies for Minority Education

In order to provide a comprehensive picture of China's tradition of positive and preferential policies for minority education, in this section I examine the theory and practice behind such policies in three cases: in imperial China, in Republican China, and in CCP-controlled areas before the founding of the PRC. I hope to make clear that, in the past half-century, the PRC government has continued the traditional practices.

Preferential and Positive Policies in Education in Late Imperial China, 1279–1911

For over two thousand years, imperial China generally took two approaches to frontier communities or other populations considered by the court and elites at the empire's center as culturally different: assimilation and accommodation (Zhou 2003, 2–8). These approaches determined what policies the imperial court created for these communities. Positive policies in education were generally made for those communities targeted for assimilation, whereas other communities targeted for accommodation usually did not enjoy the privileges of these positive policies until the imperial courts considered them ready for assimilation. Thus, imperial China's positive policies were selectively practiced. The Tang Dynasty (618–907) was probably the first imperial government to enact official positive policies when it made a connection between education and the civil service examination. The Tang's National University (*guozixue*) and local schools (*junxue*) accepted many students from frontier communities and gave them special preparation for the examination (see Lin 1990). However, the Tang policies were not systematic, either in their application or classification of groups designated for special treatment. The combination of a state framework for ethnic categorization with policies for positive or preferential treatment based on those categories did not appear until a few centuries later.

The earliest preferential treatment, based on what we would now call "ethnicity," in education and the civil examination is found during the Yuan Dynasty (1279–1368), which officially categorized or segregated people into four ethnic groups for preferential and discriminative purposes: the Mongols, the Semu (Muslim Turks and others from central Asia), the northern Chinese, and the southern Chinese. It gave preferential treatment to the ruling group, the Mongols, and to its allies, the Semu (very much like Western colonial practices during the first half of the twentieth century), but it discriminated against the Chinese, particularly the southern Chinese. For example, the National University (*guozixue*) gave half of its quotas to the Mongols and the other half to the Semus and Chinese when it reopened in 1287 (Qiao 2000, 501–502). In addition, the Yuan established Mongol and Muslim National Universities (*mengu guozixue* and *huihui guozixue*), equivalents of today's Central University for Nationalities in Beijing, which enrolled Mongols and Muslim Turks exclusively, with only a few northern but no southern Chinese. The civil examination was also divided into tests for Mongols and Semus and tests for the Chinese. The first two tests were given proportionally higher quotas for the passing grade and were much easier than those for the Chinese (pp. 540–543). For the purpose of segregation, obviously, the Yuan produced an unprecedented system of preferential and discriminatory policies and colonial practices in education and language use in China.

The Ming Dynasty (1368–1644) developed positive policies and frontier schools, the nature of which is still very much in debate (see Cai 2001, 246–249). After the Ming started to replace local chieftains with civil-service magistrates in southwestern China, its first emperor personally approved giving special quotas to indigenous chieftains so that their children could study at the National University (*guozijian*) in the capital (first in Nanjing and later in Beijing). At the same time, the ministry of interior affairs and the local governments also opened special schools for local indigenous aristocracies as part of their civilizing project. These schools waived admission examinations and provided examination-free advances to higher levels of schools (pp. 284–288). The Ming Dynasty implemented these policies mainly in southwestern China where it took an assimilationist approach to frontier communities, and such practices should be considered positive. These policies and practices created more opportunities for those indigenous students who were at a disadvantage in the regular competition for admission to the national and official local schools that prepared candidates for the civil service examination. However, it may be the case that the government actually intended to hold those children of indigenous chieftains as hostages so that the chieftains would not rebel against the central government. The Ming Dynasty might have had both intentions.

Under the Manchu rulers, the Qing Dynasty (1644–1911) started with preferential policies before it adopted positive policies. As a reward for those who fought against the Ming, it first set up various schools for Manchus, Mongols, and other northern allies, schools that were designed to give privileges to their offspring. These policies and practices were preferential and segregating only. However, motivated first by winning the hearts of the Ming loyalists, and later by Confucian-based cultural universalism, the Qing began to formulate positive policies in the late 1600s and early 1700s for the education of populations at the peripheries of Chinese civilization. This led to the development of two types of schools, *miaoxue* (Miao schools) and *yixue* (public-assisted schools; Ma 2000, 189–193, 276–287). The Miao schools originally opened for Miao and other indigenous students in Guizhou in 1659, but later enrolled indigenous students throughout the southwest. Miao schools were not always independent schools but sometimes were merely *separate classes* in regular public schools. Officials established special quotas for students enrolled in these classes to give them access to civil service examinations. Starting in the mid-1700s, the quotas would be gradually decreased as a given frontier community's Chinese proficiency rose. By the early 1800s, a limited number of quotas were still available, but those quotas might be left unfilled if enough qualified candidates were not available. As a part of the Qing's civilizing project, the public-assisted schools were not narrowly positive for only certain ethnic groups but broadly positive, which opened their doors to both indigenous students and poor, local Han students (for a case study of *yixue* in Yunnan, see Rowe 1994). This school system was more flexible in its medium of instruction, using Chinese and/or other languages, but in general its students did not enjoy special quotas for the civil service examination. Special quotas were given only in special cases. For example, after Xinjiang became a province in 1884, its governor, Liu Jintang, asked the Qing government to grant special quotas to Turkic-language speakers who had made significant progress in learning Chinese and in the study of Confucianism in public-assisted schools (Chen et al. 1998, 230–231).

As modern colleges rose in the early twentieth century, the Qing government, in 1908, established the Manchu and Mongol (and Tibetan) Language College (*manmengwen gaodeng xuetang*), which enrolled only speakers fluent in Manchu, Mongol, and Tibetan who had already passed their preliminary candidacy examinations in Chinese and their native languages. The college supported its regular students financially and offered a rather modern curriculum, including studies of Manchu, Mongol, and Tibetan languages (for the complete curriculum, see Xu 1981, 822–828). During the Qing, frontier education was taken care of by the "Barbarian" Affairs Management Office (*lifan yuan*), which, for example, prohibited the Mongolian aristocracy from hiring Chinese teachers and using Chinese for official documents until 1910 because the office feared direct communication between the two groups (Inner Mongolia 1995, vol. 2, 117–118). In 1911, however, the office proposed positive regulations (*mengzanghui difang xingxue zhangcheng*) for Mongolian, Tibetan, and Turkic Muslim groups, regulations that favored those groups in school systems, administration, teacher training, teaching materials, funding, and curricula (for the document, see pp. 118–121). Of course, the Qing Dynasty collapsed that same year and so the regulations were never implemented. The making of these regulations marked the pinnacle of over a thousand years of imperial positive policies for frontier peoples and development of school systems for these peoples, but not the end of these policies and systems.

In conclusion, in imperial China a dynasty made positive or preferential policies with two different goals. When a dynasty was ruled by the Han majority, such as the Tang (618–907), Song (960–1279), and Ming (1368–1644), the goals of its policies were generally to promote cultural universalism/imperialism and to assimilate frontier peoples into the mainstream culture and society. These policies were usually positive. When a dynasty was ruled by peoples who conquered the Han and occupied the empire's center, such as the Mongols (the Yuan 1279–1368) and the Manchus (the Qing 1644–1911), the policy objectives were often twofold: (1) to strengthen and maintain the ruling group's superior position in education, cultural affairs, and government, and (2) following the Han cultural and political mode, to assimilate more marginal peoples into the mainstream culture and society. These policies were preferential and segregating in the first case but positive in the second.

Positive Policies for Minorities in Education during the Republican Period, 1912–49

The Qing practice was continued and even strengthened during the Republican period, though civil wars and the Japanese invasion seriously disrupted policy implementations. In this section, I look into the policies of the early Republican government, the Nationalist Party's government, and the CCP within its base areas, with the understanding that Soviet influence began to reach China soon after the Bolshevik revolution in 1917.

Special consideration of minority education was on the Republican government's agenda from the very beginning. When Cai Yuanpei, as its first minister, organized the new Ministry of Education in early 1912, he designated an office of the department of regular education (*putong jiaoyu si diwu ke*) to oversee Mongolian, Tibetan, and Turkic Muslim education. During the first national conference on education, held by the Republican government in 1912, a draft of policies for Mongolian, Tibetan, and Turkic Muslim education was preliminarily passed and was forwarded to the

Ministry of Education, where it was approved (Xu 1981, 304). As a follow-up, in 1913, the Ministry published the *Regulation for Mongolian and Tibetan Schools* (*Mengzang xuexiao zhangcheng*; pp. 829–831). The regulation specified the expansion of the minority school system modeled on Qing frontier schools and specified quotas for various ethnic groups in these schools [50 percent for Mongols, 15 for Tibetans, 10 for (Turkic) Muslims, and 25 for Manchus and Hans], full financial support for students, funding sources, and curricula. The ministry's 1919 national plan for education asked for special state funding to support schools in economically underdeveloped minority communities (Song and Zhang 2005, 575–576; Xu 1981, 266).

After its reorganization in January 1924, the Nationalist Party (*guomindang*) officially adopted a policy of alliance with the Soviet Union and the CCP, a policy that was totally abandoned when the Nationalists and the Communists split in 1927 (Tung 1964, 91–111). However, the Nationalist Party's cooperation with the Soviet Union appears to have left two legacies that could have influenced China's minority education system. The first is the *Huangpu* (Whampoa) Military Academy approach for training a cadre force in a special school or class, an approach that was later extensively utilized by the Nationalist Party (Strauss 2002). This approach to education is characterized by the provision of crash courses to train students for specific military or political purposes instead of a general citizen education. The second was the development of a "party-centered" or "partified" education (*danghua jiaoyu*), a political tool of the party that taught party doctrines and was directly administered by the party (Culp 2002). Both approaches began to influence the overall Nationalist education policies after the party gained control over most of China in the late 1920s and early 1930s, but their direct impact on the Nationalist positive policies remains unclear today. The Nationalist government took a series of actions between 1929 and 1935 to make positive policies and develop schools and classes to serve minority interests, at a time when the Soviet positive policies began to decline (Martin 2001, 405–409; Simon 1991, 55). These policies encompassed positive treatment for minority students for college enrollment, extra support for schools in minority communities, and establishment of infrastructures for minority education.

To strengthen the ties between the country's center and frontiers, the Nationalist Party in 1929 adopted a decision on Mongolia and Tibet (*Guanyu mengzang zhi jueyi*), which included positive policies for minority education. Following this decision, in early 1930, the Ministry of Education held the second national conference on education, which passed a plan (*Shishi mengzang jiaoyu jihua*) for Mongolian, Tibetan, and Turkic Muslim communities (for the document, see Inner Mongolia 1995, vol. 1, 142–151).

Starting in fall 1930, the plan called for the admission of minority students by all colleges and schools, the offering of minority classes at the National Central University and Beijing University, the selection of candidate students by the relevant authorities, and financial support for minority students. Three months after the conference, the plan was written into the Ministry of Education's regulation for minority education (for the document, see pp. 130–131). In the regulation there are two points deserving special attention.

First, the term "class" (*ban*) used in the regulation may have been associated with the *huangpu* approach, though college-preparatory schools and special classes had existed for non-Han students since the Qing. It is not certain whether this was merely a change in terminology or whether it indicated an intention to use the *huangpu* approach. This question requires further study beyond the scope of this chapter.

Second, the regulation stated for the first time that Han students were not allowed to enroll in these classes, while the late Qing and early Republican regulations either focused on students' ability in Mongol and Tibetan languages, or gave more preference to minority students while not excluding Han students. However, no evidence is available to suggest any Soviet influence on this novel restriction in the regulation, and later policies show more convincingly that the Nationalist government's approach to minority students was not influenced by the Soviet one.

A few years later, in 1935, the Ministry of Education amended the regulation to allow minority students (later called "frontier students") from Xinjiang, Qinghai, Gansu, Ningxia, and Xikang to study at both public and private colleges in the national and provincial capitals (for the document, see pp. 129–130). The Mongolian and Tibetan Affairs Commission issued a regulation regarding the processes for selecting candidate students from minority communities (for the document, see pp. 134–135). Article 4 of the regulation stated that a selected candidate student must be a resident of a minority region. The residency requirement suggests that the policy gave preference to minority students who had limited Chinese proficiency and fewer opportunities rather than to "Sinicized" minority students. The spirit of Article 4 was further elaborated when the Ministry of Education published the *Regulation of Frontier Student Benefits* with nine articles (*Bianjiang xuesheng daiyu banfa*) in 1947 (for the document, see pp. 135–137). It is the last positive policy for minority students put forth by the Nationalist government, which used the term "frontier" geographically, politically, and culturally/linguistically (see Guo 1955). The positive treatment included special quotas, special financial support, lenient admission examinations, and special academic and linguistic assistance. Most importantly, the Nationalist government did not include a student in the quota simply because he or she was a member of a minority group. It took into consideration a candidate's residency in the indigenous community, linguistic and cultural difference, and financial needs. The combination of considerations as a whole was significantly different from what was practiced in the Soviet Union then and what has been practiced in the PRC since 1949.

Moreover, the 1930 minority education plan also set aside study-abroad quotas for minority students for the next eight years. Specifically, Inner Mongolia was to select ten candidates, Outer Mongolia eight, Qinghai two, Tibet eight, Xikang four, and Xinjiang four, totaling thirty-six candidate students. Selected students were required to be graduates of secondary school or above, and would study abroad for three–seven years with full financial support from the central government. However, there is no evidence to indicate that either the Soviet Union or the PRC has ever given such special consideration to minority communities. Over the years the PRC has given study-abroad quotas to Xinjiang, Inner Mongolia, Tibet, and other minority regions, but has not gone as far as to specify the ethnicity of candidates for those quotas.

In short, the Nationalist government developed a whole package of positive policies for minorities in education, and a system of frontier schools that accommodated both minority and Han students in the 1930s and 1940s. There is no evidence to suggest direct Soviet influence. In fact, my analysis of the Nationalist policies shows that the Nationalist approach was significantly different from the group approach of the Soviets and the CCP because the underlying principles were different. The Nationalist approach was based on Dr. Sun Yat-sen's Three People's Principles, seeking equality explicitly among *individuals* of all races or ethnic groups, rather than among ethnic groups per se (Inner Mongolia 1995, vol. 1, 121–123). Thus, the Nationalist positive policies were oriented toward those who were minorities and needed special assistance, not just toward those who could simply claim a minority identity. Based on the idea of individual equality, poor Han students with few opportunities were also entitled to special assistance.

The CCP's Positive Policies and Practice during the Republican Era

The CCP developed a series of minority policies before it came into power in 1949 (see Connor 1984; 67–87; Dreyer 1976, 63–92; Zhou 2003, 37–40). The CCP's earliest minority (language) education policy appeared in 1931 when the Central Chinese Soviet in *Xiang-Gan-Min* border areas in southeastern China passed its *Resolution on the National Question in*

China (*Guanyu Zhongguo jingnei shaoshu minzu wenti de juedian*; China 1981, 18–21). This document suggested the necessity of minority language education in resolving the national question, but its approach obviously copied that of the Soviet Union (for the Soviet minority school system, see Lewis 1972, 193–198). There is no evidence that the Central Chinese Soviet in southern China, some of the earliest areas occupied by CCP forces, was able to carry out the policy in the areas it controlled, probably because of the continuing civil war. Moreover, in a larger sense the absence of efforts to implement this policy was in line with the CCP's general education policy during this period, which was in full effect during the war against Japanese invasion (1937–45). The general policy mandated that cadre education had top priority and citizen education was secondary, an approach that continued the *huangpu* tradition cultivated by the Soviet advisers (Yu 2000, 292–296). The essence was fully represented in a resolution made by the CCP Politburo in Mao'ergai, in northwest Sichuan, during the Red Army's long march north in 1935 (China 1981, 33–34):

> The Red Army Political Department . . . must select some excellent minority members for class-struggle education and minority education so that they become minority cadres. After the Red Army main force reaches Shaanxi, Gansu, Qinghai, and Ningxia, more efforts along this line should be made in Hui and Mongolian communities.

The CCP began to implement this policy after the Red Army reached Yan'an in Shaanxi Province. In 1937, the CCP founded Shanbei College (*Shanbei gongxue*), which included a division training minority students for the war with Japan. At the same time, the Yan'an government also opened a Hui and Mongolian primary school in Dingbian County, the first CCP-controlled primary school to use minority languages as the medium of instruction (China 1981, 56–57). These two institutional measures during the Yan'an period (1935–48) mark the earliest CCP efforts in minority education.

In 1940, in response to Japanese efforts to set up puppet Mongolian and Muslim governments, the CCP Secretariat disseminated two important documents, the *Outline of (the Solution to) the Muslim Question* (*Guanyu huihui minzu wenti de tigang*) and the *Outline of (the Solution to) the Mongolian Question in the War against Japanese Invasion* (*Guanyu kangzhan zhong menggu minzu wenti tigang*), both of which extensively addressed positive policies for minority education (for the complete documents, see pp. 61–84). However, neither document appears to have asked for preferential treatment for the Muslims and Mongols, though the later Mongol-related document mentioned free schooling, already a common practice in CCP-controlled areas. Instead both documents required that the Muslim/Mongolian military and administrative authorities accord all peoples equal rights. These documents appear to have followed the Three People's Principles rather than the Leninist-Stalinist principle. The former stressed equality among individuals while the latter stressed equality among national groups. This may indeed be the case because the Sixth Plenum of the CCP's Sixth Central Committee made a political resolution (*Zhonggong kuoda de liuzhong quanhui zhengzhi jueyi*) to form a united front with the Nationalist government. According to the resolution, the CCP would submit its governments and armed forces to the guidance of the Three People's Principles and the leadership of Chairman Chiang Kai-shek in November 1938 (see pp. 53–55). Although its sincerity has always been questioned, at least the CCP did rhetorically follow the Three People's Principles in its public documents and propaganda work.

Following the publication of these two documents, the CCP founded the Yan'an Minority College in September 1941, modeled on the minority division of Yan'an's Shanbei College (Ha 1991, 3; Ha and Teng 2001, 421–422). The college had advanced classes (one year for students with higher education and/or leading positions), regular classes (one semester), cultural classes (for students with little education), classes for Mongolians, and classes for Muslims. The college had a total enrollment of about three hundred students from nearly ten minority groups in the early years and trained a core cadre force for the CCP before it merged with other schools in 1947. The CCP implemented similar education policies in the Mongolian and Korean communities under its control in the mid- and late 1940s.

Clearly, during the 1920s and early 1930s the CCP developed a positive policy for minority education following the Soviet model, but then during the late 1930s and 1940s, the CCP advocated equal treatment of all peoples in education rather than preferential treatment for minorities, as shown in its Mongolian and Muslim documents. This was probably done because the CCP was either bound by its cooperation with the Nationalist government (1938–45) or was at war with it (during pre-1938 years and between 1946 and 1949). Equal treatment might have been a better tool for a strong united front than preferential treatment, which might have divided communities during the years of the civil war with the Nationalists.

Summary of China's Historical Legacy for Minority Education

To implement positive and preferential policies, the governments of imperial and Republican China developed minority schools (residential and nonresidential) and classes within mainstream schools, budgeted special funding for these schools, and established quotas for minority students for places in elementary, secondary, and tertiary schools, and even in schools abroad, and provided financial assistance to them. Imperial China adopted positive policies for integration or preferential policies for segregation. When positive policies were made for integration purposes, quotas and other assistance often decreased as minority communities became more assimilated into the mainstream society. When they were intended for segregation to discourage assimilation, the preferential policies were generally maintained or sometimes intensified even as the goals were achieved. Republican China made positive policies essentially for integration purposes. Consequently, it gave positive treatment only to minorities that were to be integrated while giving minority individuals who were already integrated the same status as the Han. Clearly, imperial and Republican China considered the needs of minority communities and individuals when meeting such needs furthered the state's goals.

The PRC's Theory and Practice of Positive Policies for Minority Education

I first investigate the PRC's theoretical foundation for its positive policies, and then show how this theoretical foundation reveals continuities and discontinuities with earlier policies in the imperial and Republican eras. I particularly address the question of the Soviet Union's influence on the PRC's positive policies.

The Theoretical Foundation of the PRC's Positive Policies

The PRC's minority policies generally are based on Marxist and Leninist principles (see chapter one; Connor 1984; Dreyer 1976, 43–60), and so are its specific positive policies. Two Marxist and Leninist approaches to minorities are of particular relevance to the discussion here.

First, Marxism and Leninism take a class/group approach to any social issue. The most fundamental is the class approach that divides a society horizontally into different classes. Social conflicts are generally class conflicts, as Marx and Engels claimed in *The Communist Manifesto* (see Selsam et al. 1970, 43–51). However, when handling minority issues, which are called "the national question" in Marxism and Leninism, the Marxist and Leninist approach divides a society vertically into different national or ethnic groups, some of which may be identified as oppressors, and others as oppressed (see Lenin 1967, vol. 3, 749). These two divisions are philosophically incompatible (Connor 1984, 1), and thus politically complicated when it comes to making and implementing policy. On the one hand, national, or ethnic, conflicts should normally be subsumed under the solution to class conflicts. On the other, national conflicts may be strategically handled in the process of class conflicts, which means that the former may sometimes be given more priority than the latter. The national question in Marxism/Leninism essentially became one of how to handle appropriately the relationship between class conflicts and national or ethnic conflicts.

The CCP's treatment of self-determination and autonomy is a good example for illuminating its strategy and group approach. The CCP considered the national question one of the basic questions of the Chinese revolution (see Liu 1996, 81–88). During the years of struggle for the control of China, the CCP supported national liberation (even self-determination) as a part of the Chinese revolution, whereas, during the PRC period, the CCP has maintained the unity of all nations within its territory and opposed national self-determination. It has proposed territorial autonomy for minorities as its basic approach to the national question, as suggested by Stalin. Though administratively autonomous governments range from the *xiang* level (a level under a county) to the provincial level (autonomous regions), the approach provides autonomy to every minority community regardless of its geographic extent and population size. This is done in order to facilitate equality among minority groups, not among individuals. The group approach is also characterized by the proportional representation of minority communities in PRC political institutions. Zhou Enlai made it clear in 1950 that minority territorial autonomy facilitates the PRC's territorial and national unity, and minorities should at least be proportionally represented in autonomous governments and perhaps even represented in numbers larger than strict proportionality would require (China 1994a, 49–50).

Second, Leninism and Stalinism acknowledge actual inequality among national groups in a socialist society and consider it a "heritage" from a capitalist society, but recognize that a socialist state has a responsibility to address the inequality problem. Lenin believed that during capitalism "different nations are advancing in the same historical direction, but by very different zigzags and bypaths," and that some nations are more cultured than others (Lenin 1967, vol. 1, 172). The more cultured nations "must make up for the inequality which obtains in actual practice" (Lenin 1967, vol. 3, 749). Stalin further elaborated the point, saying that "a socialist state must eradicate the existence of actual inequality by rendering economic, political and cultural assistance to the backward nationalities" (Stalin 1935, 101).

The CCP leadership echoed Lenin and Stalin on this issue, but it was more apologetic. As early as 1950, Zhou Enlai, premier of the PRC government (1949–76), acknowledged that historically the Han wronged minorities and should apologize to them (China 1994a, 48). He further asked Han officials to embrace the attitude that they were repaying a debt in assisting minority communities (pp. 152–153). Mao Zedong, chairman of the CCP (1943–76), told a Tibetan delegation in 1952 that "If the CCP could not facilitate the development of your population, economy and culture, then it does no good" (p. 86). Liu Shaoqi, president of the PRC (1956–67), also stressed in 1954 that the Han had an obligation to provide sincere assistance to economic and cultural development in minorities communities (pp. 109–110).

The Marxist-Leninist group approach to social issues and the socialist state's obligation to eradicate inequality due to "economic, political and cultural backwardness" in minority communities has determined the PRC's group approach to the national question. The PRC's positive policies, framed by the group approach, are seen as an integral part of the solution to the national question.

The Practice of the PRC's Positive Policies for Minority Education

The CCP's principles in making and implementing positive policies for minority education have generally been followed in the PRC, though there were ups and downs during the years of political turmoil, such as the Anti-Rightist Movement (1957–58) and the Cultural Revolution (1966–76). In late October 1949, right after the founding of the PRC, Mao Zedong cabled the CCP Northwestern Bureau that all levels of government should set quotas in proportion to the populations of minority groups, recruit large numbers of members from minority groups, and train them in cadre classes or schools (pp. 42–43).

Mao's cable set the tone for the PRC's positive policies in three political dimensions—a *huangpu* approach to minority education, a proportional representation of minority groups in education, and a special quota system to ensure the former two. The *huangpu* approach, which focuses on the training of a cadre force, is a legacy from the cooperation in the 1920s among the Nationalist government, the CCP, and the Soviet Union. This legacy was continued by the CCP during the Yan'an period and then in 1950 was positioned to apply throughout the entire country. Mao considered the *huangpu* model to be an approach to the final solution to the national question because it could train numerous communist cadres from minority communities and completely isolate anticommunist reactionaries in those communities (p. 42). Mao regarded minority cadre training and territorial autonomy as two key projects at the beginning of the PRC (p. 69).

A year later, in November 1950, after Mao's instruction, the Chinese government published two documents that have had lasting significance for its positive policies from 1950 to the present (Liu 1996, 313). The first document is the *Preliminary Plan for the Training of Minority Cadres* (*Peiyang shaoshu minzu ganbu shixing fangan*), which covered eight aspects of the PRC's minority-school system, including quotas, special funding for the system, student financial aid, and medium of instruction (see Li 1981, 456–457; Liu 2000, 258–259). It became the foundation of the PRC's positive policies for minority education because all subsequent policies in the last five decades have been designed to supplement it and/or to develop it to meet emerging needs in minority education. The second document, *Preliminary Plan for the Founding of the Central Institute for Nationalities* (*Chouban zhongyan minzu xueyuan shixing fangan*), laid out the goals, means, curricula, organization, and supervision for this minority university (Li 1981, 458–459; Liu 2000, 260–261). The plan established the PRC's only model of tertiary schooling for minorities for the next five decades, a model that has become controversial in China's marketization process since the late 1990s, but was legally strengthened during the revision of the autonomy law in 2001 (see Teng 1998; Wang and Chen 2001, 313).

Of the three political dimensions that Mao specified in 1949, the special quota system for minorities is both an enduring principle and a landmark for the PRC's positive policies. It was included in the *Preliminary Plan for the Training of Minority Cadres,* and further reaffirmed and elaborated in the PRC's relevant official documents in 1951, 1956, 1962, 1978, almost every year in the 1980s, and occasionally in the 1990s and the 2000s. This does not mean that the landmark positive policy was not challenged within the CCP and the PRC government. Actually the reaffirmations were either intended to stop any deviation from the principle underlining the policy or to correct any departure from this principle. For instance, the 1962 reaffirmation was a reaction to putting too much stress on class conflict while downplaying the national question. The result was the restoration of the special quota system as a priority in the policies of the CCP and the PRC government. Similarly, the 1978 reaffirmation revived the policy of special quotas for minorities, along with the college admission examinations, both of which had been completely abandoned during the Cultural Revolution (1966–76). In the 1990s and the early years of the new century, however, the challenges to this policy have come mainly from outside the CCP and the PRC government, namely, from the forces of globalization and marketization, as shown in chapters four, five, and ten.

Of Mao's three political dimensions, however, the concept of a proportional representation of minority students in higher education was conspicuously missing in the PRC's relevant official documents from the 1950s to the 1970s, though the idea had been frequently applied to the political representation of minority groups in the legislature and government at various levels. It was not until the Third National Conference on Minority Education in February 1980 that the concept was overtly reintroduced to minority education. The conference concluded that higher education should give full consideration to college applicants' entrance examination scores, given their percentage in the total population, and special enrollment quotas. Further, efforts should be made to recruit minority students, at least in proportion to the population of minority communities in autonomous regions (China 1991, 227). Local governments expressed concerns about this document. They complained about the difficulties of enforcing a proportional representation in higher education. In response, the Ministry of Education and State Commission on Ethnic Affairs compromised and allowed this policy to be implemented according to local situations so long as the percentage of minority students would increase year by year.

Local resistance may explain why the concept of a proportional representation of minorities in higher education did not become official in the PRC's positive policies in its first three decades. The implementation of the concept of a proportional representation might help to eliminate the inherited, factual inequality among ethnic groups. However, minority groups that already had a good proportional representation in higher education would not receive quotas as large as those that did not have such a representation. Therefore, unequal allocation of quotas and financial assistance among minority groups would be seen as a violation of the Marxist principle of equality among minority groups. Moreover, the Han majority might consider some minority groups underprepared to fulfill the quotas and proportions. In short, the Leninist and Stalinist principle of equality among ethnic groups creates a dilemma in that the same positive policy has to be applied to each and every minority group under a socialist state. Regardless of the actual situations, selective application of this policy would otherwise create ethnic conflicts and conflicts between minority groups and the state. For example, if the groups receiving few quotas are powerful and strategic, their dissatisfaction may lead to political problems. The same might happen if the policy is selectively applied to some members (such as those residing in remote areas) and not others (such as those residing in prosperous, urban communities) of the same group. Doing so would not only violate the equality principle but also infringe upon the group approach, disturbing unity within the minority group. For reasons having to do with principle and the reality of ethnic politics, the PRC government has practiced its positive policies for minority education without the flexibility found in imperial and Republican practices, though they all shared the same inventory of measures, such as a system of minority classes/schools, quotas, special budgets, and special financial assistance.

The Question of Soviet Influence

I have narrowed the Soviet influence down to theory mainly, much narrower than suggested in existing scholarship on this subject (Dreyer 2006, 284; Sautman 1999). This case can be made more convincing by further examination of direct Soviet influence on the PRC's minority policies from two perspectives, communication between the two parties' leaderships and work by Soviet advisers in China.

Stalin and the Soviet Communist Party indeed intended to influence the CCP's minority policies. In early 1949, A. I. Mikoyan, representing Stalin, visited Mao Zedong and passed along Stalin's message that the CCP should not allow minority self-determination, but give minority groups only territorial autonomy (Ledovskii 1999/2001, 85). This piece of advice might have influenced Mao and the CCP, but Liu Chun, former vice chair of the PRC State Commission on Nationalities (1952–66), suggested that the idea of territorial autonomy in China originated with Mao Zedong's speech at the Sixth Plenum of the Sixth Central Committee of the CCP in 1938 (Liu 1996, 305–306). Regardless of the origin of the idea, on October 5, 1949, the CCP Central Committee instructed its regional bureaus and field-army CCP committees that the term "self-determination" should not be used in its minority policy anymore because it might be used by imperialists and minority reactionaries to sabotage the unification of China (China 1997, vol. 1, 24–25). Further, in June 1949, Liu Shaoqi headed a CCP delegation on a secret visit to Moscow for two months. According to the memorandum of the first official talk, Stalin reaffirmed his earlier message to Mao that the CCP should increase the Han population in Xinjiang from 5 to 30 percent to achieve a solid control of that area (Ledovskii 1999/2001, 100). The CCP seems to have heeded this advice, too. The Chinese government increased Xinjiang's Han population from 6.7 percent in 1949 to 32.86 in 1964 (see China 1994b, 26). However, there is no published and unclassified evidence that the leadership of the two communist parties communicated specifically on positive policies for minorities, especially in education.

During his Moscow visit, on July 6, 1949, Liu Shaoqi presented to Stalin a letter requesting advice and advisers. The list included requests for assistance in areas ranging from the structure of the central government to specific technical support (for the letter, see Ledovskii 1999/2001, 116–120). Regarding education, the list contained requests for information on the organization of schools at various levels, the relationship between schools and students' future employers, college enrollment and student life, and secondary-school curricula. Stalin granted Liu Shaoqi most of his wishes, if not all. When he returned to China in August, Liu Shaoqi had over 200 Soviet advisers accompanying him, but probably no educational advisers (see Shen 2003, 72). In May 1950, the Soviet Union dispatched the first group of 42 educational advisers who were full professors, associate professors, and instructors, and by 1952, a total of 187 educational advisers had been sent to China (pp. 111–14). They had extensive impact on higher education in China (Kong 2004, 128–131), but there is no evidence that any of them were ethnologists or worked at the Central Institute (now University) for Nationalities and other minority schools during this period.

The Soviet ethnologists did not arrive until the mid-1950s. The first adviser in minority work was G. P. Serdyuchenko, who, along with several other Soviet linguists, arrived in Beijing in October 1954 (Zhou 2003, 169–196). He served as adviser to the Institute of Minority Languages of the Chinese Academy of Sciences, and as linguistic adviser to the Central Institute for Nationalities. He provided the Soviet model of minority-language maintenance and development for his Chinese colleagues, and much of his work was technical and involved standardizing writing systems for minority languages. His policy-oriented advice, which was accepted by the State Commission on Nationalities, was to ally writing systems (using the Cyrillic script) across Sino-Soviet borders and to ally writing systems across minority communities within China. However, both alliances were abandoned in 1957 when domestic politics changed and Sino-Soviet relationships deteriorated.

The first Soviet ethnologist, N. N. Cheboksarove, came to Beijing in July 1956, though the work of Soviet ethnologists had already begun to change Chinese ethnology by then. As adviser to the president of the Central Institute for Nationalities, Cheboksarove taught graduate courses, gave a series of lectures on ethnology, and helped the institute make its research plans for the 1956–57 and 1957–58 academic years (Wang et al. 1998, 94–105). He contributed significantly to Chinese scholars' understanding of Soviet and Western ethnology.

However, after nearly ten years of archival research in China, I have not found any evidence suggesting direct Soviet influence on the PRC's positive policies for minority education. It does not appear that the leaders of the two parties communicated on this subject. The PRC government had already formulated its positive policies for minority education and had started to implement them by the time the Soviet ethnologists arrived in Beijing. In fact, some evidence suggests that the Chinese government intentionally fended off direct Soviet influence in its minority affairs. For example, the PRC State Commission on Nationality Affairs and the Ministry of Foreign Affairs were cabinet-level organs that had no Soviet advisers (Shen 2003, 110; Zhou's personal communication with officials of this State Commission, 2006). This is a crucial issue because the head of the Soviet advisory group working within a ministry could participate in the ministry's CCP committee meetings and ministers' meetings. Policies and major decisions used to be, and are still, made at these meetings. Further, there is evidence that the PRC State Commission on Nationalities Affairs kept some distance from the Soviet ethnologists working at the Central University for Nationalities. For example, after his arrival in Beijing, Cheboksarove wanted to survey the State Commission's policy making/implementing process and prepared a questionnaire with a few dozen questions for the commission to answer. The commission passed the questionnaire all the way to Ulanhu, then vice premier of the PRC and chair of the commission, who instructed to "ignore it" (Zhou's personal communication with officials of the State Commission, 2006).

The lack of evidence of direct influence, however, may not be proof of the absence of indirect influence from the Soviet Union. My earlier discussion shows how Mao Zedong's 1949 cable shaped the development of a system of minority schools in China. It is possible that Mao only wanted to strengthen a development started in Yan'an when minority schools and classes were established by the CCP. Yet, it is also possible that Mao was inspired by the Soviet system of minority schools. At a CCP Politburo meeting on nationalities affairs in July 1953, Mao commented on the importance of a strong system of minority colleges, saying, "We need a Party school for minorities. The Soviet Union has the University of the Toilers of the East while we have the University of the West (of China)"—meaning the Central Institute (University) of Nationalities in Beijing (Liu 1996, 315–316). It is not clear whether Mao meant that China modeled itself after the Soviet Union, or that China was comparable to the Soviet Union in training minority cadres, or only that China should match the Soviet Union in this area.

Summary of the PRC's Positive Policies and Practices

Marxism, Leninism, and Stalinism were the guiding ideology for the PRC when it was founded in 1949. As such, they provided the guiding principles in the PRC's policy making, including the making of positive policies. Marxists and Leninists recognize the existence of factual inequality among ethnic groups in a socialist state, but blame it on the preceding capitalist societies. They take a class/group approach to social issues, an approach underlining how the PRC has sought national equality. This became the theoretical foundation for the PRC's positive policies. The Soviet Union was the model for the PRC's interpretation of Marxism and Leninism. It is only in this sense that the Soviet Union may have directly influenced the PRC's positive policies. No evidence available so far suggests any direct influence by the Soviet leadership or Soviet advisers on the PRC's practice of its positive policies in the 1950s.

Conclusions

The survey presented earlier shows that imperial and Republican China had a system of schools (residential and nonresidential) and classes within regular schools for peoples identified as outside the sphere of the Chinese civilization. There were special budgets for these schools, quotas for their students to study in elementary, secondary, and tertiary schools and even abroad, and financial assistance to them. Following what is now called a "need-based" principle, imperial and Republican China practiced the positive treatment with great flexibility, constantly making adjustments in accordance with the goals of the state and the perceived needs of minority communities, though the preferential policies and practices were discriminative and segregating.

The PRC government used to advantage many of these earlier measures in its own positive practices. However, the flexibility of these earlier measures was compromised by the Marxist-Leninist group approach to national equality borrowed from the Soviet Union. This principle not only constrains the PRC government's flexibility in its practice of positive policies, but also causes two major problems.

First, the group approach emphasizes equality of groups while it ignores inequality of groups. For example, when it provides special quotas for college enrollment, it usually provides them to all groups, regardless of a group's proportional representation in the college student population. It is difficult, if not impossible, for the PRC government to make policies targeting some minority groups with underrepresentation at the college level while excluding some minority groups with near- or above-proportional representation. In many cases, the CCP and PRC government actually practice what I call a Hobbesian principle of equality: Minority groups posing greater threats to the PRC territorial and national integrity benefit more from the principle of group equality than do minority groups posing lesser challenges, thus perpetuating inequality (Zhou 2004).

Second, the group approach focuses on equality among groups while it ignores inequality within a group. Measuring equality or inequality against the majority group, the PRC's policies blindly provide positive treatment to a minority group. For example, larger quotas go to urban minority residents than rural minority residents, who have more difficulties than the former. The PRC government is fully aware of these problems. However, it does not want to violate the Marxist and Leninist principle as long as it remains the foundation of its positive policies. Moreover, the historical guilt on the part of the majority also makes the Han-dominated government hesitant to withdraw the benefits of its positive policies from minority communities whose students outperform average Han students. These problems show how the underlying Marxist and Leninist principle denies the PRC government the flexibility that imperial and Republican China enjoyed in practicing positive policies. To help minorities that really need help, the PRC government must abandon the Marxist and Leninist group approach that has been practiced in the form of a Hobbesian principle of equality, and adopt a need-based approach to both individual and group equality. Thus, it could practice its policies with the greater flexibility to meet the needs of minority individuals and groups. After the collapse of the Soviet Union in 1992, the CCP and the PRC government have gradually replaced the Soviet model of multinational state-building with a Chinese model of one nation with diversity (Zhou forthcoming). This Chinese model has given the PRC government more flexibility in the making and implementation of positive policies (see chapters five and ten). However, a path to equality for minorities eventually lies in a political approach (democratization) to group equality as well as in a legal approach to individual equality, an approach that, ironically, Lenin criticized in 1920 (Lenin 1967, 422–423).

References

Cai, Shoufu. 2001. *Yunnan jiaoyu shi* [History of Education in Yunnan]. Kunming: Yunnan jiaoyu chubanshe.

Chen, Shengyuan, Wenhua Ma, Zongzheng Xue, and Tayierjiang. 1998. *Weiwu'erzu jiaoyu shi* [History of Uygur Education]. In *Zhongguo shaoshu minzu jiaoyu* [History of Minority Education in China], ed. Han Da, vol. 1, 185–455. Guangzhou: Guangzhou jiaoyu chubanshe.

China. 1981. *Dang de minzu zhengce wenxian ziliao xuanbian,1922.7–1949. 10* [Selected Documents of the Chinese Communist Party on Minorities Policy: July 1922–October 1949]. Beijing: Zhonguo shehui kexueyuan minzu yanjiusuo.

———. 1991. *Shaoshu minzu jiaoyu gongzuo wenjian xuanbian 1949–1988* [Selected Documents of the Central Government on Minority Education: 1949–1988]. Hohhot: Neimengguo jiaoyu chubanshe.

———. 1994a. *Zhongguo Gongchandang zhuyao lingdaoren lun minzu wenti* [Chinese Communist Party Leaders on the National Question]. Beijing: Minzu chubanshe.

———. 1994b. *The Population of China Towards the 21st Century, Xinjiang Volume.* Beijing: Zhongguo tongji chubanshe.

———. 1997. *Jianguo yilai zhongyao wenjian xuanbian* [Selected Important Documents since the Founding of the PRC], 20 vols. Beijing: Zhongyang wenxian chubanshe.

Connor, Walker. 1984. *The National Question in Marxist-Leninist Theory and Strategy.* Princeton, NJ: Princeton University Press.

Culp, Robert. 2002. "Setting the Sheet of Loose Sand: Conceptions of Society and Citizenship in Nanjing Decade Party Doctrine and Civics Textbooks," in *Defining Modernity: Guomindang Rhetorics of a New China, 1920–1970*, ed. Terry Bodenhorn, 45–90. Ann Arbor, MI: The Center for Chinese Studies, University of Michigan.

Dreyer, J. Teufel. 1976. *China's Forty Millions.* Cambridge, MA: Harvard University Press.

———. 2006. *China's Political System: Modernization and Tradition.* New York: Pearson.

Guo, Lianfeng. 1955. "Bianjiang jiaoyu" [Frontier Education], in *Zhonghua minguo jiaoyu zhi (er)* [History of Education in the Republic of China, II], ed. Wu Junsheng, 1–35. Taibei: Zhonghua wenhua shiye chuban weiyuanhui.

Ha, Jingxiong. 1991. *Zhongguo shaoshu minzu gaodeng jiaoyuxue* [Studies of Minority Higher Education in China]. Nanning: Guangxi minzu chubanshe.

Ha, Jingxiong, and Xing Teng. 2001. *Minzu jiaoyuxue tonglun* [Introduction to Minority Education]. Beijing: Jiaoyu kexue chubanshe.

Inner Mongolia. 1995. *Neimenggu jiaoyu shizhi ziliao* [Historical Data on Education in Inner Mongolia], 2 vols. Hohhot: Neimengguo daxue.

Kong, Hanbing. 2004. *ZhongSu guanxi jiqi dui Zhongguo shehui fazhan de yingxiang* [Sino-Soviet Relationship and Its Impact on the Social Development in China]. Beijing: Zhongguo guoji guangbo chubanshe.

Ledovskii, Andrei Mefodievich. 1999/2001. *Sidalin yu Zhongguo* [Stalin and China]. Trans. Chunhua Chen and Zunkuang Liu. Beijing: Xinhua chubanshe.

Lenin, Vladimir Il'ich. 1967. *V. I. Lenin Selected Works*, vol. 3. New York: International Publishers.

Lewis, Glyn. 1972. *Multilingualism in the Soviet Union: Aspects of Language Policy and Its Implementation.* The Hague: Mouton.

Li, Weihan. 1981. *Tongyi zhangxian wenti yu minzu wenti* [United Front Issues and the National Question]. Beijing: Renmin chubanshe.

Lin, Liping. 1990. "Sui Tang de bianjiang zhengce" [The Sui and Tang Dynasties' Policies for Frontier Regions], in *Zhongguo gudai bianjiang zhengce yanjiu* [Studies of Imperial China's Policies for Frontier Areas], ed. Ma Dazheng, Lin Ronggui, and Liu Di, 150–181. Beijing: Shehui kexue chubanshe.

Liu, Chun. 1996. *Liu Chun minzu wenti wenji* [Selected Works of Liu Chun on the National Question]. Beijing: Minzu chubanshe.

———. 2000. *Liu Chun minzu wenti wenji (xuji)* [Selected Works of Liu Chun on the National Question, vol. 2]. Beijing: Minzu chubanshe.

Ma, Yong. 2000. "Zhongguo jiaoyu zhidu tongshi, Qingdai (shang) (Gongyuan 1644–1840)" [History of Educational Systems in China: the Qing from 1644 to 1840], in *Zhongguo jiaoyu zhidu tongshi* [History of Educational Systems in China], ed. Li Guojun and Wang Bingzhao, vol. 5. Jinan: Shandong jiaoyu chubanshe.

Mackerras, Colin. 1994. *China's Minorities: Integration and Modernization in the Twentieth Century*. Hong Kong: Oxford University Press.

Martin, Terry R. 2001. *The Affirmative Action Empire: Nations and Nationalism in the Soviet Union, 1923–1939*. Ithaca and London: Cornell University Press.

Ogden, D., and M. Perelman (eds.). 1960. *Soviet State Law*. Moscow: Foreign Languages Publishing House.

Postiglione, Gerard. 1999. *China's National Minority Education: Culture, Schooling, and Development*. Now York: Falmer Press.

Qiao, Weiping. 2000. "Zhongguo jiaoyu zhidu tongshi, Song, Liao, Jin, Yuan (Gongyuan 960–1368)" [History of Educational Systems in China: the Song, Liao, Jin, and Yuan Dynasties, 960–1368], in *Zhongguo jiaoyu zhidu tongshi* [History of Educational Systems in China], ed. Li Guojun and Wang Bingzhao, vol. 3. Jinan: Shandong jiaoyu chubanshe.

Rowe, William T. 1994. "Education and Empire in Southwest China," in *Education and Society in Late Imperial China, 1600–1900,* ed. Benjamin A. Elman and Alexander Woodside, 417–457. Berkeley, CA: University of California Press.

Sautman, Barry. 1998. "Preferential Policies for Ethnic Minorities in China: The Case of Xinjiang," *Nationalism and Ethnic Politics* 4 (1 & 2): 86–118.

———. 1999. "Expanding Access to Higher Education for China's National Minorities: Policies of Preferential Admissions," in *China's National Minority Education: Culture, Schooling, and Development*, ed. Gerard A. Postiglione, 173–210. New York: Falmer Press.

Selsam, Howard, David Coldway, and Harry Martel (eds.). 1970. *Dynamics of Social Change: A Reader in Marxist Social Science from the Writings of Marx, Engels, and Lenin*. New York: International Publishers.

Shen, Zhihua. 2003. *Sulian zhuanjia zai Zhongguo, 1948–1960* [Soviet Advisers in China, 1948–1960]. Beijing: Zhongguo guoji guangbo chubanshe.

Simon, Gerhard. 1991. *Nationalism and Policy towards the Nationalities in the Soviet Union: From Totalitarian Dictatorship to Post-Stalinist Society*. Trans. Karen Foster and Oswald Forster. Boulder, CO: Westview Press.

Song, Enrong, and Han Zhang. 2005. *Zhonghua minguo jiaoyu faguai xianbian* [Selected Educational Laws and Regulations of the Republic of China], revised edn. Nanjing: Jiangsu jiaoyu chubanshe.

Stalin, Joseph. 1935. *Marxism and the National and Colonial Question*. New York: International Publishers.

Strauss, Julia C. 2002. "Strategies of Guomindang Institution Building: Rhetoric and Implementation in Wartime Xunlian," in *Defining Modernity: Guomindang Rhetorics of a New China, 1920–1970*, ed. Terry Bodenhorn, 195–221. Ann Arbor, MI: The Center for Chinese Studies, University of Michigan.

Teng, Xing. 1998. "Ershiyi shiji woguo minzu xueyan de banxue moshi ying zuo zhongda diaozheng" [Our Country Should Adjust Its Model of Minority Higher Education for the Twenty-First Century], unpublished manuscript of a speech at the Minority College Presidents' Conference.

Tung, William L. 1964. *The Political Institutions of Modern China*. The Hague: Marinus Nijhoff.

Wang, Geliu, and Jianyue Chen. 2001. *Minzu quyu zizhidu de fazhuan: Minzu quyu zizhifa xiugai wenti yanjiu* [The Development of Autonomy for Minority Regions: Studies on the Issues in the Revision of the Law of Regional Autonomy for Minority Nationalities]. Beijing: Minzu chubanshe.

Wang, Jianmin, Zhang Haiyang, and Hu Hongbao. 1998. *The History of Ethnology in China*. Kunming: Yunnan jiaoyu chubanshe.

Xu, Xincheng. 1981. *Zhongguo jindai jiaoyu shi ziliao* [Archival Data for the History of Education in Modern China]. Beijing: Renmin jiaoyu chubanshe.

Yu, Shusheng. 2000. "*Minguo shiqi (Gongyuan 1912 zhi 1949)*" [The Republican Period (1912–1949), in *Zhongguo jiaoyu zhidu tongshi* [History of Educational Systems in China], ed. Li Guojun and Wang Bingzhao, vol. 7. Jinan: Shandong jiaoyu chubanshe.

Zhou, Minglang. 2003. *Multilingualism in China: The Politics of Writing Reforms for Minority Languages 1949–2002*. Berlin/New York: Mouton de Gruyter.

———. 2004. "Minority Language Policy in China: Equality in Theory and Inequality in Practice," in *Language Policy in China: Theory and practice since 1949*, ed. Minglang Zhou and Hongkai Sun, 71–95. Amsterdam/New York: Kluwer.

———. Forthcoming. "The Fate of the Soviet Model of Multinational State-Building in China," in *China Learns from the Soviet Union, 1949–Present*, ed. Tom Bernstein and Hua-yu Li. Lanham, MD: Rowman & Littlefield Publishers.

Chapter 3

Preferential Policies for Minority College Admission in China: Recent Developments, Necessity, and Impact

Tiezhi Wang

Chinese colleges and universities have adopted preferential policies to lower their cutoff points for ethnic and linguistic minority college applicants (see chapters two and four). Involving the interest of millions of people in China, these policies have had a broad impact and have received extensive scrutiny, generating criticism and controversy. This chapter analyzes how these policies have evolved, why they have been made and how they have been recently adjusted, and what they have achieved in the last half-century in China.

Social Functions of China's National College Admission Examinations

To grasp the meaning of China's preferential policies for minority college applicants, it is necessary to have a comprehensive picture of the social functions of China's national college admission examinations. The national college admission examinations began in 1952 (Zhang 1984, 337) and have since then been administered, as the academic criteria for college admission, to college applicants throughout China, except during the interruption of the Cultural Revolution (1966–76).

The first function of the national college admission examinations is to select the best talent. It is the best way to utilize limited higher educational resources so that college education opportunities are given only to those who are the most competent students. Usually those who have higher examination scores are more competent students with greater potential. Thus, the education system relies on examinations to evaluate students and identify the best. The second function of the national college admission examination is to distribute limited educational resources fairly. It is socially more progressive to distribute educational opportunities according to academic talent rather than according to social status or financial means. When college education is still not mass education, it seems most equitable to adhere to the examination score equality principle that every college applicant is equal when s/he is judged by his/her college admission examination scores.

However, we are aware that to disperse educational resources only on the basis of an examination score is not a perfectly equitable system. There are ethnic and regional disparities in educational resources and quality of local school systems in China. If only one cutoff score is adopted nationally, applicants from some minority communities and in some regions will lose their opportunities to go to college, resulting in inequality in access to college education. Thus, China has created some preferential policies to redress the inequality in higher education access while maintaining the practice of unified national college admissions based on admission examination scores. With regional differences from one province to another, these preferential policies cover a wide range of beneficiaries, including ethnic minority applicants, Han applicants from poverty-stricken areas, overseas applicants, military veteran applicants, and so on (for details, see Wang 2007a).

Recent Developments in Preferential Policies for Minority College Applicants

China's preferential policies for minority college applicants started in the early 1950s when the Ministry of Education stipulated that "college applicants who are workers, government employees, and soldiers with three years or more experience, as well as minorities and those from overseas should be shown lenience in the admissions process when their examination scores are lower than the cutoffs" (Wu 1998, 209). This policy changed very little until China's economic reforms started in the late 1970s (for details, see chapters four and six). Since then China has developed a variety of preferential policies for minority college applicants to redress new situations brought about by China's economic and political reforms. These policies mainly target minority applicants from border areas, mountainous areas, and pastoral areas and have lower college admission cutoff scores for them in accordance with educational development levels in their home communities. There is also a policy that gives priority to minority applicants over Han applicants when these minority applicants live in Han communities. Preferential policies also mandate lowered cutoff scores for college applicants from specially targeted minority communities and/or for minority applicants committing to return to work in targeted areas upon graduation (for details, see Wang 2001, 231).

Provinces used to have the authority to decide how many points for minority applicants were to be deducted from the cutoff scores for regular admission. However, in 2002, the Ministry of Education for the first time required that for minority applicants from border, mountainous, and pastoral area, no more than twenty points may be deducted from the local cutoff scores (MOE 2002). Because there are discrepancies among minority areas, depending on their distance from larger urban areas, provinces and autonomous regions have adopted a wide variety of measures since then.[1] The flexibility is usually based on ethnicity, regional economic development, languages of instruction and/or language of the admission examinations, and other factors.

In some provinces/regions, cutoff scores are lowered based on applicants' ethnicity. In Ningxia Hui Autonomous Region, Hui applicants receive twenty bonus points while other minority applicants in the same region are given ten bonus points. In Xinjiang, when taking the admission examinations in Chinese, college applicants from Uygur, Kazak, Mongolian, Kirgiz, Tajik, Xibe, Uzbeck, Tatar, Daur, Tibetan, and Russian groups are given fifty bonus points if both parents belong to one of these groups, but only ten points if only one parent belongs to these groups. The same ethnic group may enjoy different bonus points in different provinces. For example, applicants from Daur, Oroqen, and Ewenki groups receive ten bonus points if they live in Inner Mongolia, but twenty if they reside in Heilongjiang Province. In Tibet, there are different cutoff points for Tibetan and Han students, a gap that is sometimes close to one hundred points.

In other provinces or regions, different bonus points are given to minority applicants from different areas, taking into consideration the autonomous status of some minority areas and economic development. In Hebei Province, for example, minority applicants from autonomous counties get twenty bonus points, while those who live in Han communities receive only ten. In Sichuan, minority applicants from its three autonomous prefectures and seven autonomous counties are given fifty bonus points, those from concentrated minority communities in urban areas get ten points, but those minority applicants from Han neighborhoods do not receive any bonus points. In Guangxi, minority applicants from major urban areas, such as Nanning, Guilin, and Liuzhou, enjoy only five bonus points while those in rural areas are given twenty points. Yunnan Province adopts preferential policies based on levels of economic development and proximity to China's borders (for details, see chapter five). It should be pointed out that the consideration of economic development in a region also favors, as well, Han students from poverty-stricken and underdeveloped areas. Ningxia, Gansu, Qinghai, Yunnan, and Tibet all began to give bonus points to Han applicants within their jurisdictions in the last decade. These provinces/regions use different standards in the actual practice of their preferential policies. For example, Gansu gives ten bonus points to Han applicants from underdeveloped areas whereas minority applicants receive twenty points. Qinghai rewards Han applicants up to twenty bonus points if they have lived and gone to school locally for ten or more years.

Some provinces take language of instruction and/or language of the college admission examinations into consideration in implementing their preferential policies. Qinghai Province in western China gives thirty-five bonus points to minority college applicants who take the admission examinations in Chinese and are residents in its six autonomous prefectures. Liaoning Province in northeastern China awards ten bonus points to those minority students who have trilingual education (mother tongue, Chinese, and English) in school while giving only five to those who do no have it. Also in northeastern China, Jilin Province awards ten points to minority applicants who take an admission examination in their native languages. In Xinjiang and Inner Mongolia there are different cutoff scores for applicants whose high school instruction media are minority languages versus those who are taught in Chinese. For example, in Inner Mongolia in 2005, the cutoff scores for Chinese medium applicants were 525 for humanities majors and 555 for sciences majors, but the cutoff scores for minority language media applicants were 409 for humanities majors and 466 for sciences majors.

There are also preferential policies for specially designated minority classes in colleges and universities as well as for preparatory classes (see chapters four and five). The Ministry of Education stipulated in 2003 that the maximum bonus points for minority classes could be as many as forty, for four-year college preparatory classes, eighty, and for two-year college preparatory classes, sixty (MOE 2003). In addition, as early as 1984, graduate schools in China also began to practice preferential policies in admissions. These policies allow bonus points for graduate school applicants who are from border, mountainous, and pastoral minority communities. The actual practice varies from year to year. For example, in 2006 the Ministry of Education authorized graduate schools to give thirty–seventy bonus points to minority applicants' oral admission examination if they successfully passed the written admission examination. Also in 2006 the Ministry of Education began a minority "advanced talent" training program for the coming years (MOE 2006). This program gives minority graduate school applicants national admission examinations and special cutoff scores with bonus points.

It is clear that the Chinese government has taken an active approach to redress historical and new inequalities in access to college and graduate school education for minorities, in general, and for Hans in poor areas.

Preferential College Admission Policies and Ethnic Equality

Preferential policies are needed for minority students with lower college admission examination scores. However, it is crucial to analyze why minority applicants have lower scores if we truly want to understand the necessity of preferential policies for educational equality.

Lower academic achievements usually reflect levels of individual talent and motivation, as well as family background, schools' teaching quality, and the social environment. Within the latter three factors often lies the source of the academic gap among different ethnic groups. International scholars have put forth theories to account for the academic gap, and these theories may help us understand the differences in academic achievement among ethnic groups in China (Ha and Teng 2001, 57–65). For example, the cultural deprivation theory hypothesizes that due to the lack of family and community learning activities and stimuli found in mainstream white schools, black children in the United States do not have as strong a motivation, ambition, and linguistic/cognitive ability as white children, and thus have lower academic achievement. To take another example, language style theory posits that the difference between family language and school language leads to lower academic achievements because such difference hinders communication between students and teachers. Cultural conflict theory assumes that minority students' learning style, values, attitudes, and behavior cultivated in their families and communities often come in conflict with the mainstream campus culture, resulting in minority students' lower academic achievements (see chapter nine). Taking advantage of these theories in analyzing lower academic achievement among China's minority students, I examine this phenomenon from the following three perspectives:

First, social, economic, and cultural factors have significant impacts on minority education and minority students' academic achievement. Economically most of China's minority communities are located in underdeveloped areas in western China where local governments rely on the central government for funds, on the average, for about 35 percent of their annual budgets (Hu 2001, 306). These local governments do not invest enough in education (see chapter five). For example, in 2005 educational funding for each elementary, middle, and high school student was 2,075, 2,650, and 5,942 RMB, respectively, in eastern China, while it was only 987,

1,165, and 2,823 RMB per student, respectively, in western China (Ketizu 2005). Per student educational funding at the elementary, middle, and high school levels in western China was not even half of that in eastern China. As for livelihood, the population in western China mainly relies on agriculture (63.5 percent) with smaller proportions working in industry (12 percent) and services (24.5 percent) (Long 2004, 257). This underdevelopment limits the job market and local residents' views of education, science, and technology. All these have a negative impact on minority students' motivation in school. Culturally, school language (*Putonghua*/Chinese) is usually different from family and community languages in western China. Although progress has been made in bilingual education, the lack of qualified bilingual teachers and materials still adds to the difficulties that minority students face in schools. For instance, when I was doing fieldwork in Deang communities in Yunnan between 2000 and 2001, I found that without a written language Deang children had to read Chinese textbooks and immerse in Chinese language instruction as soon as they started their primary school (Wang 2007b, 95–99). They still did as well as their Han classmates did at the very beginning, but they gradually lagged behind because language became an increasingly formidable barrier. The language barrier leads to poor academic performance among minority students and puts them at a considerable disadvantage in local and national examinations.

Second, schools are one of the most influential factors in minority students' academic achievements and eventually in their national college admission examinations. In China, key secondary schools have the best teachers, the most modern equipment and libraries, and plenty of financial resources. Up to 90 percent of their graduates may pass national college admission examinations. However, these key schools are always located in economically well-developed towns in eastern China and only in provincial and prefectural capitals in western China. Minority students typically have no access to those key schools unless they happen to live in those urban areas. The farther away from cities and economic centers, the fewer resources a school has. The fewer resources a school has, the less it prepares its students academically for the national college admission examinations. Minority students usually go to these poor schools. As a result, even after all the efforts the Chinese government has made, fewer minority college applicants successfully pass the examinations and go to college. According to the 2000 national census data, in the Han majority, there are 357 persons with college education for every 10,000, but the Blang, Va, Lisu, Lahu, Hani, and Deang groups in Yunnan produce only 49, 41, 49, 47, 64, and 43 college graduates per 10,000 people, respectively (China 2004, 124–183). No members of the Deang group have had the opportunity to go to graduate school yet.

Third, families influence their children's education in two ways: financially and culturally/cognitively. In China there are more minority families under the poverty line than well-to-do ones. In 2003, 3.1 percent of China's rural population was under the official poverty line, but 7.3 percent of the minority population was in poverty. Of the seventy-seven poverty-stricken counties where minorities live, there were 3,900,000 persons in poverty, about 24 percent of the local population (Bianxiezu 2006, 48–49). These poor minority families cannot financially support their children's education. Some minority students drop out of school because their families cannot pay for fees, board, and textbooks. Some minority students cannot concentrate their efforts on their studies because they have to labor in the fields for their families. Moreover, minority parents' lack of education also has a negative impact on their children. In China, the education levels of minorities are generally lower than those of the Han, with the exception of a few groups such as the Koreans and the Manchu. According to the 1990 national census data, the national average illiteracy rate was 22 percent, but 61–74 percent of the Dongxiang, Monba, Lhoba, and Tibetan minorities were illiterate and 30–57 percent of the populations of the Hani, Li, Lisu, Va, Lahu, Blang, Salar, Primi, Nu, Deang, and Bonan were also illiterate (see Zhang 1998, 273; Zhou 2000). These illiterate parents cannot help their children with their studies even if they want to do so. Their children lack home-cultivated cognitive strategies to handle school work successfully, and cannot get academic help at home. They are financially and cognitively disadvantaged.

To redress the three negative factors mentioned earlier, China's preferential policies for minority college admission are the key measures to guarantee minority applicants equal access to higher education. By lowering the cutoff scores, giving minority applicants bonus scores, and admitting them into remedial classes, these policies tackle the three negative factors, singly or collectively. By lowering the national college admission examination cutoff scores for different minority regions, the government tries to solve the problem created by the uneven distribution of financial and educational resources. For example, in 2004, the cutoff scores were 600 for humanities majors in the coastal Shandong Province but in Tibet were 490 for the Han and 350 for Tibetans (Gaokao 2004). The difference in cutoff scores between Shandong and Tibet was targeted at closing the regional gap in resources. If the regional difference had not been taken into consideration, even Han college applicants in Tibet could not have been admitted into college, not to mention the Tibetan applicants. The different cutoff scores for the Han and Tibetan applicants in Tibet were not solely based on ethnicity, but also took into account differences in access to resources. The Han applicants generally came from urban centers in Tibet, while the Tibetan applicants were usually from rural and pastoral areas that lacked educational resources. These preferential policies attempted to put students from different socioeconomic and educational backgrounds on equal footing in the competition for college education.

China's national college admission examinations are usually culturally (and linguistically) biased against minority applicants but favor Han applicants. For example, examinations on Chinese history often focus on the history of the Han Chinese. Moreover, a composition topic, whether addressed in Chinese or minority languages, may be related to themes of urban and modern life, which Han students experience more often than minority students do. If these examinations were based on a minority culture and history, Han students would do poorly, too (Xie 1986). Lower cutoffs and bonus points are given to offset the cultural and linguistic bias in the examinations.

When well-controlled, the lowered cutoffs and bonus points do not necessarily affect the quality of education for minority and majority students in college. First, for universities that admit students from throughout China, it is common to find examination score differences as high as forty points, even when lowered admission scores for minority students are not considered. This score variation is natural among a large and diversified student population. Bonus points provided by preferential policies are often restricted to within this variation range so that policy beneficiaries are sufficiently prepared for the academic challenge. Student applicants needing bonus points beyond this range are unlikely to succeed. Second, lowered cutoff scores are usually set for local universities that belong to lower tiers and admit lower tier students as well. Their goals are not to train top, talented students on a national level, but students who can meet local needs. Third, when lower cutoff scores benefit minority students who are not well prepared for key universities, they are generally sent to preparatory or remedial classes opened by these universities. In general, China's preferential policies for college admission do not affect the quality of college education when they are well implemented.

Assessment of the Impact of China's Preferential Policies for College Admission

China's preferential policies for minority college admission have a positive impact on three areas—financial cost, social cost, and symbolic significance—with few side effects.

First, the policies have increased cultural and linguistic diversity at colleges and universities in China at a minimal financial cost. Unlike other preferential policies for minorities, such as tuition and fee waivers and aids to minorities in poverty-stricken areas, preferential policies targeting minority enrollment at colleges do not bring any financial burden to the Chinese government. These policies simply redistribute a small portion of the existing opportunities for college education from the Han majority, who, as a group, are minimally affected, to minorities. In a word, these policies are easy to implement without requesting a lot of financial resources.

Second, these preferential policies significantly increase minorities' access to college education with little social cost. China's minority population is currently under 9 percent, which means only a small number of minority students relative to China's total population can enjoy the benefits of the preferential policies. The implementation of these policies does not significantly affect Han college applicants' opportunities, but significantly increases opportunities for minority college applicants. According to statistics from 2005 college applications in Gansu Province in western China (Li 2006), out of over two hundred thousand applicants, only about eighty-five hundred were the potential beneficiaries of preferential policies for minorities, comprising about 4 percent of the total applicants. In the end, the actual beneficiaries of these policies were an even smaller percentage of the total number of applicants. However, the bonus point policies allowed the percentage of admitted minority applicants relative to the total population of minorities in Gansu to approach that of the admitted Han applicants, which has ranged from 40 to 50 percent nationwide in the last few years. Thanks to these preferential policies, minority college applicants find more access to college education while Han applicants do not see a significant decrease in such access, particularly as China has been expanding enrollments in its universities since the end of 1990s.

Third, these preferential policies are of great symbolic significance for ethnic equality and the cohesion of the Chinese nation. These policies target all minorities in China as potential beneficiaries for equal opportunity and equal access to college education, though the actual beneficiaries are a small number of minority college students. When a minority college student benefits from these policies, the impact is felt by his or her family and community, who, thus, see equality and find belonging in China.

Of course, as with any social policy, the preferential policies for minority college admission have some negative consequences too. One of the most obvious is for the individual, since these policies are based on ethnic groups as a whole. The result is that sometimes when two college applicants, one a Han student and the other a minority student, are compared, the Han applicant may feel some unfairness in certain situations (see chapter four). For example, when the minority applicant's college admission examination scores are the same as or slightly lower than those of the Han applicant, the minority applicant may be admitted to a college, while the Han applicant is not, or the minority student may be accepted to a higher ranked college than the one that accepts the Han student. Then, the Han applicant will complain about the fairness of these preferential policies. Another problem is that some college applicants take a free ride on the benefits of these policies. They change their ethnic identities to enjoy the benefits, claiming to be minority when they have any kinship connection to a minority ancestry. Some college applicants go even further to fabricate minority identities for these benefits. These applicants take away opportunities from minority applicants who are the intended beneficiaries of the preferential policies, and thus cause damage to these policies. A third problem is the perceived negative impact that the policies have on minority students' motivation to learn, but more research is needed to explore if there is indeed such an impact and how extensive it is if it exists.

Conclusion and Suggestions

My discussion shows that China's preferential policies for minority college admission are morally justifiable and practically effective. These policies have increased opportunities for minority college applicants, sped up the training for needed minority talents, helped to achieve educational equality for all ethnic groups, and facilitated ethnic cohesion in China.

However, recent local policy adjustments for social, economic, and regional factors may have decreased the scope and intensity of China's earlier preferential policies based solely on ethnicity. The new policies benefit more applicants now, but not necessarily minority applicants, as statistics suggests. Between 1990 and 2000, China's minority population increased from 8.01 to 8.41 percent, but during the same period the number of minority college students decreased from 6.6 to 5.8 percent of the total college student population in China (China 2004, 487 and 516). The decrease is directly related to the adjustments in preferential policies for college admission, though other factors may be involved. Further research is needed to identify the impact of these adjustments.

I strongly believe that, to ensure equal access to college education for all ethnic groups, China should continue its ethnicity-based preferential policies and improve the efficiency of these policies. There are four aspects of these polices that currently deserve more attention.

First, awareness should be raised among policy makers that ethnicity-based preferential policies cannot be replaced with region-based ones. It is reasonable to consider the regional factor in economic development and educational resources, but these factors should not be used to exclude or weaken the ethnicity factor in policy making or adjustment. Even within the same poverty-stricken area there is a significant difference between the local Han and minorities in educational development, culture, and language. The national college admission examinations are still biased, linguistically and culturally, against minority applicants, while favoring local Han applicants. It is morally wrong and politically damaging to give up ethnicity-based preferential policies.

Second, ethnicity-based preferential policies should be adjusted to account for differences among minority groups. More benefits should be given to those minority groups whose educational development is significantly lagging behind and whose cultures and languages are greatly different from the mainstream culture and Chinese language. In addition to college admission, preferential policies should also be deployed to improve basic education, remedial education, and local higher education so that the problems in equal access to education can be fundamentally resolved.

Third, preferential policies should address new concerns such as equal access to education for minority migrants. Recently, as China's migrant population reached over 150 million, many members of minority groups have also joined this wave of migration to urban and coastal China. They bring with them their cultural and linguistic difficulties and their educational deficit. Policy makers and implementers should study how to help these minority migrants gain access to education as they migrate.

Fourth, relevant departments of the government should regularly scrutinize how preferential policies are implemented in schools. As China's college admissions gradually move from a strict quota system to quota- and market-driven parallel admission systems, universities now have more flexibility in their admissions. In the state's quota system universities enroll students within an assigned quota, but in the market driven system universities enroll any student who can pay full tuition, regardless of their ethnicity. This also means that universities have more flexibility in implementing or not implementing preferential policies for minority applicants. Universities may take advantage of the flexibility to benefit financially at the expense of minority applicants (see chapter four). Local governments should regularly check how well the central government's preferential policies are implemented in local universities and correct any problems in minority admission and retention in a timely fashion.

Note

1. The cutoffs and bonus points for minority applicants come from provincial college admission bulletins. For details and updates, visit http://www.eol.cn and the college admission website for each province and autonomous region.

References

Bianxiezu. 2006. *Zhongyang minzu gongzuo huiyi jingshen xuexi fudao duben* [A Tutorial Reader of the Essentials of the Central Government's Conference on Minority Work]. Beijing: Minzu chubanshe.

China. 2004. *Zhongguo minzu tongji nianjian* 2004 [2004 Annals of China's Ethnic Minority Statistics]. Beijing: Minzu chubanshe.

Gaokao. 2004. 2004 *nian quanguo gedi gaokao luqu fenshu xian* [2004 College Admission Cutoff Scores in Various Provinces/Regions in China]. http://gaokao.chsi.com.cn/gkxx/fsx/2004.shtml [in Chinese].

Ha, Jingxiong, and Xing Teng. 2001. *Minzu jiaoyuxue tonglun* [An Introduction to Minority Education]. Beijing: Jiaoyu kexue chubanshe.

Hu, Angang (ed.). 2001. *Diqu fazhan—xibu kaifa xin zhanlue* [Regional Development: New Strategies for Opening up Western China]. Beijing: Zhongguo jihua chubanshe.

Ketizu. 2005. "Suxiao chaju: Zhongguo jiaoyu zhengce de dongda mingti" [Narrow the Gap: A Crucial Issue for China's Educational Policies], *Beijing shifan daxue xuebao* 5: 5–15.

Li, Xueping. 2006. "Gansu gaokao benke luqu zuidi kongzhi fenshuxian huading" [Gansu Province's decision on the lowest cutoff scores for college admission]. *Lanzhou chenbao*, June 26.

Long, Yuanwei (ed.). 2004. *Zhongguo shaoshu minzu jingji daolun* [An Introduction to Economy in Ethnic Minority Areas in China]. Beijing: Minzu chubanshe.

Ministry of Education (MOE). 2002. *Gaodeng yuanxiao zhaosheng gongzuo guiding* [Regulations for College Admission]. http://www.eol.cn/article/20020227/3038948.shtml [in Chinese].

———. 2003. *Guanyu quanguo putong gaodeng xuexiao minzu yukeban, minzuban zhaosheng, guanli deng wenti de tongzhi* [Notice on the Admission and Management of Remedial and Special Classes for Ethnic Minorities]. http://www.moe.edu.cn [in Chinese].

———. 2006. *Peiyang shaoshu minzu gaocengci rencai jihua de shishi fangan* [Plans for Implementing the Program for Training Advanced Talents for Ethnic Minorities]. http://www.lawbook.com.cn/law/law_view.asp?id=99139 [in Chinese].

Wu, Shimin. 1998. *Zhongguo minzu zhengce duben* [A Reader of China's Minority Policies]. Beijing: Zhongyang minzu daxue chubanshe.

Wang, Tiezhi. 2001. *Xinshiqi minzu zhengce de lilun yu shijian* [Theory and Practice of China's Minority Policies During the New Era]. Beijing: Minzhu chubanshe.

———. 2007a. "Preferential Policies for Ethnic Minority Students in China's College/University Admission," *Asian Ethnicity* 8 (2): 149–163.

———. 2007b. *Deangzu jingji fazhan yu shehui bianqian* [Economic Development and Social Changes in the Deang Community]. Beijing: Minzu chubanshe.

Xie, Xiatian. 1986. "Yucexing ceyan fenshu de shuangchong hanyi" [The Double Meanings of Examination Scores]. *Guangming ribao*, December 12.

Zhang, Tianlu. 1998. *Minzu renkouxue* [A Study of Ethnic Minority Population]. Bejing: Zhongguo renkou chubanshe.

Zhang, Jian (ed.). 1984. *Zhongguo jiaoyu nianjian* [Annals of Education in China]. Beijing: Zhongguo dabaikequanshu chubanshe.

Zhou, Minglang. 2000. "Language Policy and Illiteracy in Ethnic Minority Communities in China," *Journal of Multilingual and Multicultural Development* 21 (2): 129–148.

Chapter 4

Preferential Policies for Ethnic Minorities and Educational Equality in Higher Education in China

Xing Teng and Xiaoyi Ma

Historically, there have been great socioeconomic disparities and inequalities between China's minorities and the majority Han. As has been discussed in chapters one through three, after the People's Republic of China (PRC) was founded in 1949, influenced by Marxist national theories, the Chinese government instituted preferential policies for minorities to improve their access to education (for principles for policy making, see chapter two; for specifics on Marxism, see chapter one). However, with China's transition to a market economy from a centrally planned economy beginning in the 1980s, preferential policies on access to higher education have generated challenges and controversies.

In this chapter, with a brief review of the development of the PRC's preferential policies for higher education, we analyze the reality of current higher education for minorities in the context of the U.S. affirmative action policies and their rationales, and we critically question China's current preferential policies for higher education, focusing on three issues: (1) conflicting views on the preferential policy to lower the bar on college admission examination scores for minorities; (2) controversies about preparatory (remedial) classes; and (3) the difficulties for minority students from underdeveloped regions as a result of tuition and other expenses related to higher education. In our conclusion, we provide some suggestions and solutions for equality for minorities in higher education in China.

The Development of Preferential Policies for Higher Education within the PRC

Preferential College Admission for Minority Applicants

Minority education in China developed late in the 1950s when many minority communities did not have a modern education system. The minority population was then mainly composed of illiterate and semiliterate persons. In some minority communities education made rapid progress, but it was still not up to the standards of education in China's interior. Implementing preferential measures for minority students was intended to provide more opportunities for them to advance up the education hierarchy and go on to college.

Regarding the preferential treatment for minority students on college admission examinations, as early as 1950, the PRC government for the first time stipulated that "though their examination scores are a little lower, brother minority students can be shown leniency in admissions" (see Zhaosheng 2005). This regulation was in effect for three years. From 1953 to 1961, the regulation was revised so that minority students were given "priority on admission when having the same score as Han applicants" (Zhaosheng 2005).

In 1956, PRC's Ministry of Higher Education conferred and issued *A Notice on Preferential College Admission for Minorities*. The document stressed that "Considering the special conditions of minority students, more admission opportunities should be given to them. As long as their course grades can reach the lowest admission standard, preferential admission should be given to them" (China 1991, 218–219). However, this policy was neglected during the Great Leap Forward (1957–58) when it was assumed that ethnic minorities could integrate into the Han mainstream overnight.

In 1962, the central government pointed out in "Report on the Working Conference for Ethnic Minorities" that preferential admission measures should be resumed for minorities. To implement this policy, the Ministry of Education and State Commission on Nationality Affairs issued "A Notice on Preferential Admission for Minorities to Higher Education." Three articles of this new policy specifically regulated preferential admissions: (1) preference should be given to minority college applicants when they have the same admission examination scores as Han applicants; (2) more preferential considerations should be given to minority applicants if they apply for colleges in their own autonomous regions and if their examination scores reached the lowest standard set by Ministry of Education; and (3) the classical Chinese examination should be waived for minority students who apply to humanity and social science programs. However, this policy was again suspended during the Cultural Revolution (1966–76). During this period the universities were closed between 1967 and 1970 when existing students were sent to factories, farms, and army camps and no freshmen were admitted. When colleges admitted worker-soldier-peasant students between 1971 and 1976, preferential admission was given to minority students on the basis of their class status instead of their ethnicity.

In 1978, when the national college admission examination resumed after the disruption of the Cultural Revolution, the policy mandated that "the lowest admission cutoff point and specific college admission point ranges may be applied with discretion to minority applicants from border areas" (see Wen et al 2001, 585; Zhaosheng 2005). In 1980, the Ministry of Education reaffirmed three points in college admission regulations: (1) some key universities should set up preparatory/remedial classes and lower their cutoff scores in order to enroll minority applicants from border, mountainous, and pastoral areas; (2) other non-key universities can, at their discretion, show leniency in considering minority students' scores for admission; and (3) universities should give priority to minority applicants who live in Han communities when these applicants have the same scores as Han applicants (see Zhaosheng 2005). This policy was consistently practiced during the 1980s and early 1990s when the Ministry of Education systematically assigned college admission quotas to specific areas and colleges to tackle certain groups' underrepresentation in higher education.

In 1994, to implement the Chinese Government's *Zhongguo jiaoyu gaige fazhan gangyao* (China's Education Reformation and Development Program), college admission practice was adjusted to coordinate college admission on the state quota track with college admission on the locally controlled track. The latter track had been in effect since the mid-1980s (for local admission policies, see MOE 1995; for responses to the reform, see SEAC 1996, 895–954). These two tracks for college admission have complicated the practice of preferential admission for minority students. The locally controlled admission track provides more freedom for colleges to practice preferential admission but also leads to some problems. The preferential admission policies have undoubtedly helped increase minority college admission. In 1951, there were only 2,117 minority undergraduates, which was 1.36 percent of China's college student population at that time, whereas in 1996 there were as many as 196,800 minority undergraduates, who then made up 6.5 percent of the total college student population in China (see Wen 2001, 586–587).

Preparatory/Remedial Classes for Minority Students

Before the founding of the PRC, one-year preparatory classes had been set up during the Republican period (1912–49) but later cancelled (see chapter two). Once the Chinese Communist Party (CCP) took over in 1949, some preparatory classes for minority students were again established early on in minority institutes, minority universities, trade schools, adult colleges, and other post-secondary institutions. The first preparatory education classes went into effect at the Central University for Nationalities (CUN) in Beijing in 1953. The goals of the preparatory classes were to educate minority youth from remote minority areas, to help them improve their knowledge base, and to remediate their academic preparedness for four-year college study in the political science departments at CUN, other universities, and trade/professional schools. Further, at CUN flexible and effective measures were taken, including remedial classes providing two–four years of schooling at different levels, from junior high school to senior high school. In 1956, there were 1,448 preparatory students, amounting to over two-thirds of the total number of students at CUN (Song 2002). Since 1980, preparatory classes at CUN have developed into two modes: One is called "(special) Chinese classes" (*Hanyu ban or Hanyu zhuanxiu ban*) and the other "national minority preparatory classes" (*quanguo minzu ban*). The first type with two years of schooling aims to improve Chinese proficiency for minority students from Xinjiang. The second with one year of schooling enrolls minority students and prepares them academically for Tsinghua University, Beijing Normal University, CUN, and other key universities, which those qualified students may choose to enroll in directly for four-year study.

Minority preparatory classes have since spread throughout the country. Between 1980 and 1998, minority preparatory classes enrolled a total of 90,000 minority students, having prepared many qualified minority students for higher education. For example, by 1996 there were minority preparatory classes in as many as 140 universities, with total enrollments of 11,622 students. By 2001, minority enrollments in preparatory classes had increased to 13,000 (Song 2002). As a special form of education for minority students and an important part of minority education, preparatory education has been playing a significant role in broadening the knowledge base for minority students and helping more minority students receive higher education.

In March 1984, the Ministry of Education and State Commission on Nationalities Affairs made some explicit regulations for preparatory classes (see Wang 2001, 229). According to the regulations, the task of preparatory classes is to take special measures, according to minority students' needs, to improve their basic level of knowledge, reinforce their basic academic skill training, and facilitate their comprehensive development in moral, academic, and physical education, in order to lay a solid foundation for their four-year college education. The duration of schooling in the preparatory classes is generally from one year to two years, the latter usually for students with poor Chinese proficiency. After their preparatory education schooling, the students enroll in the four-year education curriculum, just as other college students do. How much to reduce scores admitting students to preparatory classes is based on factors such as training goals, requirements of different colleges, and, moreover, various minority communities' actual situations.

Problems and Challenges for Preferential Policies for Minorities in Higher Education

Recently there has been considerable debate worldwide about preferential policies for minorities in education; China is no different. After a brief look at the controversy in the United States, we examine in more detail one specific case in China to demonstrate the issues involved in that country's debates over preferential policies for minority education.

Granting ethnic minorities preferential treatment in college admissions in the United States and in other multiethnic countries has provoked controversy, both among academics and in society at large. In the 1960s, the American government implemented an "Affirmative Action Program" that aimed to eliminate discrimination in employment, education, and so on for ethnic minorities and eventually for women. Later it evolved into an explicit compensatory policy for minorities and women, giving them preferential consideration in enrollment, employment, and so on so as to compensate for their disadvantages resulting from discrimination (see Liu 2001). However, questions and debates about the "Affirmative Action Program" continue today (see chapter twelve). The focus of the higher education debate is whether minority students should enjoy preferential treatment in college admission (for details and analysis of the legal aspects of affirmative action, see chapter thirteen).

The adoption of preferential policies for college admission has become an international hot topic, as well, and in China has been controversial for the past decade. For example, at the CUN in 2000, the first author of this chapter gave to one of his seminars the following case for discussion. When carrying out a survey in Xinjiang, the author met a Han farm worker serving in the local reserve corps, who complained about the inequity of the current preferential policies. He compared his child with his Uygur neighbor's child. These two boys were born in the same year, went to the same kindergarten, primary school, and middle school in the same year, and received the same score for their college admission examination. However, the Uygur boy was matriculated at a key university in Beijing because of "bonus points" for college admission scores for minorities, while the Han farm worker's went to an inferior local university. He complained that this is unfair. Generally speaking, his son should have enjoyed the same priority because three generations of his family made contributions to the country's frontier construction in Xinjiang. The author asked his students: "Is the preferential policy for college admission fair in this case?" The students in the seminar, coming from different regions, socioeconomic statuses, and ethnic backgrounds, engaged in a heated discussion and expressed four different views.

The first view is that the preferential policy for college admission for minority applicants is a policy of prejudice and discrimination against ethnic minorities. The minority students who held this view came from nonminority and urban areas. They opposed the policy of giving bonus points for minority applicants and see it as shameful. They believe that minority applicants should get access to college education without any preferential policy and based on their own ability.

The second view is that college admission policies should consider not only the applicants' ethnic group identities but also their regions' level of economic development. Some minority and Han students who held this view thought that preferential policies should serve minorities in remote and underdeveloped minority regions, where there is a lack of basic education resources. They argued that minorities living in cities should not enjoy the preferential policy because, like their Han neighbors, they have access to the superior education resources typically found in China's urban areas.

The third view is that the preferential policy for college admission should be based on socioeconomic standing instead of ethnic group identity. They argue that some priority should be given to those college applicants who are socioeconomically disadvantaged because they have had less access to education resources than those from more affluent families.

The fourth view supports the current preferential policy for college admission. The students who expressed this view came from remote and underdeveloped minority communities. They argued that ethnic group identities are the only criterion on which the policy should be based. They cited the huge gap and inequality between minorities and the Han. They felt that minorities need preferential policies to overcome the deleterious gap.

Challenges for Preparatory Classes and the Cost of Tuition in Minority Education

Problems have gradually emerged for the preparatory class system. The major problem is that these classes enroll more children of local officials and the socioeconomic upper class than children of the disadvantaged people in minority communities. Quotas for admission to these classes are limited for any minority community. Some officials use power to get them, while others use money. These days a preparatory class student pays tuition and fees, totaling 30,000–50,000 RMB (US$5,000–7,500), and sometimes as much as 80,000–100,000 RMB (US$12,000–15,000). Thus, not only does the local quota constrain educational opportunity for minority students, but even if accepted under the quota system, minority students encounter further impediments to educational opportunity because of the high tuition and fees for the preparatory class. This situation completely undermines the original intent of the policy.

With the transition to a market-oriented economy from the state-planned economy, a tuition-based system in higher education has replaced one that in the past was nearly cost-free. Now colleges charge a minority student 2,000–5,000 RMB (US$300–750) annually for tuition and even higher for some hot majors. A student has to pay as much as 8,000–12,000 RMB (US$1,200–1,800) annually when room and board are included. Minority students from underdeveloped areas, where the annual GDP is below 1,500 RMB (US$230) per person on the average, cannot afford such an expensive college education. Going to college has caused economic troubles for many minority students and their families. Is this fair to disadvantaged minority students, or do these circumstances effectively undermine equality of access of educational resources?

Reflections on Equality for Minority Education during the Transition to a Market Economy

Preferential Policies for College Admission and Their Challenges

Since China's reforms began in the late 1970s, great changes have taken place in the historical and social circumstances against which decisions about preferential policies have been made. China has transformed from a planned economy to a market-oriented economy. The old pattern of distribution of social and economic resources for minorities has been changing while new patterns are emerging in reallocating social and economic resources to them. Thus, the intended beneficiaries of preferential policies for college education have been changing, too, because the socioeconomic status of members of minority communities changed. Objectively, the preferential policies based solely on ethnic group identities cannot meet the needs of diversified and stratified minority communities in a plural society. Subjectively, perceptions of preferential policies and attitudes toward policy beneficiaries have become more varied, as China's social groups, including ethnic groups, have changed. China's market economy has inevitably led to greater gaps between ethnic groups and among members of each group, increasing the sense of urgency among some that the government ought to take steps to ensure a fair and just system for equal opportunities. People have developed stronger and more sensitive awareness of their own individual rights and stronger commitments to the pursuit of justice for their rights. Given this new consciousness, preferential policies, if applied only to minorities, will displease others, like the Han farmer in the case given earlier, who saw their own entitlements to education threatened. The question for policy makers is: how should these policies be improved so that they will meet the needs of a changing society?

To see the need for change, we return now to the case examined earlier. *Article 9 of Education Law of the PRC* stipulates "By law, a citizen has a right to receive higher education. Citizens legally have an equal opportunity to receive an education regardless of their ethnicity, race, gender, profession, socioeconomic status, and religious belief, etc." (China 1998, 2). It is necessary that we adhere to the principle for equality in education for everyone. However, the earlier case of Xinjiang shows a conflict in equality and fairness in access to education, a conflict that legally goes against *Article 9 of the Education Law.* To some extent, the farm worker's complaint that the two boys have unequal opportunities in access to higher education is valid. However, the current preferential policy for college admission is based on ethnic group identities and aims at creating equal opportunities for education for ethnic minorities. It is beyond doubt that the Uygur boy is a legitimate beneficiary intended by the preferential policy. But the Han boy, three generations of whose family have been making their efforts to support the state's frontier construction and who shares the same disadvantaged geographic, financial, and educational environment with the Uygur boy, should also enjoy some preferential treatment, in order to realize, to some degree, equal opportunity for education for everyone.

We should note that the principle of equality in the *Education Law* is based on individual equality, while the preferential policies for minorities in higher education adhere to equality of ethnic groups and put group equality over individual equality. Our case, in fact, fully demonstrates the conflict between individual equality and group equality (see also chapter two). The key issue is how to adjust and balance the relationship between individual equality and group equality. The West (particularly the United States) puts more stress on individual equality, though group equality is sometimes considered. China's Marxist minority theory and policies have tended to emphasize group equality more, which is the result of its historical circumstances and the need to maintain ethnic group equality, political stability, and national unity. But China's current market economy demands equal opportunity for individuals and equal access to education for the individual. Are the existing preferential policies contradictory to the new demands? Is it fair for Uygur and Han children from the same neighborhood to have unequal opportunities?

In fact, the preferential policies for minorities in higher education in China are special admissions policies for minority students. These policies have been devised, in consideration of historical and practical social conditions and with the premise of full respect for ethnic differences, in order to better realize ethnic national equality. According to U.S. scholar Douglas Rae's principle of choice equality, everyone should receive equal treatment. If there is no obvious excuse, no one should be discriminated against. If there are differences between people, discriminatory treatment is needed, based on the "equality" concern. That is, to achieve genuine equality, different people should be treated in different ways (Xu 2000; also see chapter twelve). Preferential policies based on the identity of ethnic groups aim at group equality. The policies aim at narrowing the gap among individuals and eventually realizing individual equality by means of reducing group inequality for a certain period of time. As we illustrate later, achieving this strict equality of ethnic groups in practice leads to what is called "affirmative discrimination" or "reverse discrimination" undercutting the quest for educational equality among individuals.

With the transition and development of China's economy, the content and meaning of preferential policies for minorities should also adapt to meet the needs of the new situation. Taking the interests of minorities into consideration, the policies should fully embody the ideas of equality for minority students and educational fairness, and reduce political, cultural, economic, and educational gaps among ethnic groups. This goal can only be achieved by means of state-made preferential policies.

UNESCO pointed out in December 1960 that equality of educational opportunity involves two meanings: eliminating discrimination and inequality. Eliminating discrimination means that everyone enjoys equality in education whatever his or her race, color, gender, language, religion, politics, social birth, or family background. Eliminating inequality means to banish any differential treatment between regions and groups that is not the result of intentional prejudice and discrimination (Ma and Gao 1998, 86). In brief, equality in educational opportunity demands that everyone has an equal chance to receive education and achieve success in education. It is the primary motor of upward mobility in modern society and should be available to everyone, regardless of his or her social status, identity, ethnicity/race, and gender.

Coleman (1968) suggested that educational equality should be surveyed from various angles and spoke of multiple concepts of educational equality, including equality of educational opportunity, equality of educational process, equality of educational resource, equality of educational result and educational effect, as well as rectification equality and compensating equality (Xu 2000). Rectification equality means to compensate, by financial measures, those who are excellent in ability but do not have privileged backgrounds. The key point about compensating equality is to compensate those who are born in terrible environments. In 1966, in his famous Coleman Report, Coleman (1968) pointed out that students should be provided an equal admission opportunity, and, in addition, that the teaching process and the results from being educated should also be equal (Xu 2000). Equality of educational opportunity globally more and more focuses on social minorities, such as ethnic minorities, females, the disabled, and the poor. After the 1990s, equality of educational opportunity was extended to mean the whole process of education and its outcomes, the final aim of which is to ensure everyone benefits from the requirement of basic educational opportunity.

In light of research on educational equality, equality of educational opportunity is defined such that regardless of his/her race, class, ethnicity, gender, religious belief, and economic status, everyone has equal rights and opportunities to receive education. For the PRC, this means that everyone is entitled to a fixed number of years of basic compulsory education, as well as to equal and full opportunity to receive other kinds of education necessary to develop individual potential. Thus, what we stress about the equality of educational opportunity echoes Coleman's concerns: *equality of educational access, equality of educational process,* and *equality of educational result.* At present China's preferential policies for minorities in higher education mainly deal with equality in educational access. Equal rights to receive an education stipulated in *Educational Law of the PRC* also mainly aim at equality of educational access, a goal directly reflecting conditions in China. Equality in the letter of the law, whether it refers to individuals or ethnic groups, pursues absolute equality. However, given the reality of cultural and economic difficulties among China's peoples, actual equality is always relative to the particular circumstances of the individual or group.

We think that the Xinjiang case discussed earlier involves adjusting and balancing the relationship between individual equality and group equality. From the individual's perspective, the Han farm worker's complaint about reverse discrimination raises the issue of unfairness for Han college applicants caused by preferential college admission legally institutionalized. From the group's perspective, the preferential policies for college admission were established to develop education for minorities, to train talented students for work in ethnic regions, and to narrow gaps between ethnic groups with different historical backgrounds. Based on ethnic group identity and guided by group equality theory, the implementation of preferential policies in higher education aims to balance the interests of all ethnic groups, to ensure every group's economic and social rights, and to compensate for disparities arising from economic, cultural, and developmental factors in the past. The laws compensate in particular for minorities' disadvantages in competition for educational access in an environment dominated by Han mainstream culture, to which minority college applicants, such as the Uygurs, may have little exposure. We think that education inequality in this context has two important dimensions. First, there are actual differences in education levels among and within ethnic groups, as well as differences in other realms such as family income, culture, and exposure to *Putonghua*. Second, there is the perception that preferential policies for minority applicants often discriminate against Han college applicants. This perception, in addition to reflecting people's experiences with the system, may also reflect increasing awareness of the conflict between preferential policies mandating differential treatment in college admissions on the basis of ethnic group membership, and the *Education Law*, which guarantees education rights to the individual citizen. In short, the so-called education equality is only an ideal educational model. Absolute individual equality that the Han farm worker wanted in the Xinjiang case does not exist in reality. "It is really uncommon to find simple individual equality in education because education systems can never completely provide the same treatment for every student" (Xu 2000).

At present, the preferential policies for college admission are still necessary and play a significant role. Giving proper preferential consideration to minority applicants not only increases their opportunity and equal rights in education, but also facilitates the development of education for minority communities and the reduction of gaps between their communities and the mainstream. Of course we should not ignore some negative impacts from these policies in some regions and groups. Problems resulting from the lack of systematic considerations of the needs of different groups have gradually appeared during the transition to a market economy. In addition to ethnic differences, the policies should be improved to take regional differences, socioeconomic differences, and differences in basic education sources into consideration in adopting compensatory measures.

Preparatory (Remedial) Classes and Challenges

It is a fact that minority preparatory classes have been very effective as a bridge for many minority students' smooth transition to a four-year college education. However, much attention should also be paid to the negative side that preparatory classes have revealed recently. To a large extent and against the original intent of the policy, preparatory classes as an educational resource now tend to serve more children of the elite in minority communities than the children of disadvantaged families in those communities. As a result, many talented minority students from underprivileged families have no access to this resource and to college education. In the past there were loopholes in the policy that the elite could take advantage of for their children. Now, to seek more financial resources, colleges legitimately, and often illegitimately, charge higher tuitions and fees for preparatory classes, which effectively gives the opportunity only to those who can afford it. Fueled by the competition for resources in the new market economy, this situation undermines the intent of the policy for equality of educational access, leading to inequality of educational access in minority communities.

We believe that the original intent of the policy for preparatory classes is correct and helpful to equality of educational opportunity. We suggest that the implementation of this policy should strictly follow Article 71 of the *PRC's Law on Regional Autonomy for Minority Nationalities*, which stipulates that "all levels of the government and schools should adopt a variety of measures to help minority students from socioeconomically disadvantaged families to pursue their education" (China 2002, 27–28). The state should enforce the law and ensure that this preferential policy benefits the underprivileged minority students for whom it was intended.

Tuition and Fees for Minorities in Higher Education

In 1997, the tuition policies for higher education were put into practice, charging every student, including ethnic minorities, for the cost of their higher education in order to relieve colleges' shortage of financial resources. Inevitably, it came as a great shock to lower income families, particularly for those who live in underdeveloped minority regions in western China, to pay the high tuition and fees. According to the government's 2002 statistics, China's population may be divided into four socioeconomic classes: upper class, middle class, lower class, and the lowest class. China's upper class had a population close to 45 million (3.5 percent of China's population), where families had a net annual income of 20,000 RMB (about US$3,000) per person and most of whom lived in eastern China. China's middle class, with a population of 450 million (35 percent of China's total population), had a net annual income of 6,000–7,000 RMB (US$1,000) per person and were found mostly in towns in eastern and central China. China's lower class, with a population of 700 million (54 percent of China's population), had a net annual income of only 2,000 RMB (US$300) per person and lived in rural and western China. The lowest class, with a population of 100 million (80 million rural and 20 million urban), had a net annual income below 500–700 RMB (US$75–$100) per person in rural China and was an indicator of deep poverty in the urban areas. Over 60 percent of China's minority population of 106 million was categorized as lower class. Moreover, the majority of those in the lowest class were ethnic minorities (Teng 2004). Currently colleges charge a student between 5,000 and 10,000 RMB (US$740–1,500) annually for tuition, fees, and room. Almost 30 percent of college students had difficulties paying this amount (Zhao 2002). Many minority students belong to this group. Although a financial aid system, including scholarships, loans, tuition/fee-reduction, and tuition/fee-waivers, has been put in place along with the tuition/fee system, it has flaws and limited resources.

This situation adds another barrier to equality of educational access for minority students. In addition to the problem of academic preparedness and achieving higher scores for college admission, minority students face the problem of affordable higher education. The negative effect of affordability is seen not only in the lower percentage of minority students in college currently compared to the early 1990s, but also in higher dropout rates in middle school and high school (see chapter nine). Some minority families have already given up hope for a college education for their children because they know they cannot afford it even if their children are able to gain admission to a university. In an educational system where the goal is to move to higher levels of education, the loss of hope for upward mobility makes it harder to ensure enrollment, retention, and graduation rates of minority students even in primary and secondary schools. Obviously current preferential policies for college admission that ensure minorities' equal access to higher education are insufficient in a market economy. The state should adjust its policies to provide sufficient financial aid so that minority students have real equality with their peers in access to university education.

Suggestions for Preferential Policies

After thoroughly scrutinizing and reconsidering China's current preferential policies for college admission, we have the following five suggestions for readdressing the problems we've identified.

First, the principle of ethnic equality must be followed (see Zhou 2002). As long as there are cultural differences and socioeconomic stratification among ethnic groups in China, our preferential policies should follow the approach of "different but equal" treatment (see chapter twelve). At present, preferential policies should continue to be implemented for minorities because there is still a long way to go to realize true educational equality.

Second, a system should be established to balance educational resources in minority and non-minority areas. With the development of Chinese society, the existing preferential policies that have shown deficiencies in political, economic, and cultural arenas and in resource allocation must be improved. Currently, the beneficiaries of preferential policies have changed and so have their value orientations. Differences have developed among and within ethnic groups so that social stratification is undergoing reconfiguration and resources are undergoing reallocation. The socioeconomic gap between different areas and within specific ethnic groups is the main factor causing the inequality. The new system should be able to address this new problem efficiently and sufficiently.

Third, preferential policies for college admission should be revised, abandoning the criterion for eligibility solely based on ethnic group identity. The revision should also take account of the differences among individuals, regions, cultures, and socioeconomic levels (see Wang 2003). We believe that the approach of "different but equal" treatment is an ideal model in pursuit of equality in education. As long as the differences among ethnic groups exist, educational inequality will not disappear. However, we should understand that tomorrow's educational equality has to face today's educational inequality if we want to realize true equality among ethnic groups in the end.

Fourth, the preferential policy for preparatory/remedial classes should be reconsidered from two perspectives to ensure true equality for minorities in higher education. It needs to be revised to target children of unprivileged minority families, particularly those from minority communities in remote poor regions. In addition, to ensure the effectiveness of this preferential policy, the government should revise relevant educational laws and enforce them consistently.

And last but most important, with regard to tuition and fees in higher education, a preferential policy should be created for financial assistance for minority students from remote and underdeveloped regions. Just as Coleman (1968) argued, we should take financial compensatory measures for those who have excellent capability but no privileged background so as to achieve "remedial equality" and for those who are in disadvantageous environments to realize "compensatory equality." The state should set up a more effective financial assistance system in higher education. Colleges should charge tuition and fees on the basis of students' ability to pay, taking into consideration ethnic, regional, and socioeconomic differences.

References

China. 1991. *Shaoshu minzu jiaoyu gongzuo wenjian xuanbian* (1949–1988) [Collection of Policy Documents for Ethnic Minority Education, 1949–1988]. Huhehaode: Neimenggu jiaoyu chubanshe.

———. 1998. *Zhonghua Renmin Gongheguo xiangxing jiaoyu fagui huibian 1990–1995* [Collection of Current Education Laws and Regulations of the PRC, 1990–1995]. Beijing: Remin jiaoyu chubanshe.

———. 2002. *Zhongguo minzu quyu zizhi falu fagui tongdian* [Collection of Laws and Regulations on Minority Regional Autonomy in the PRC]. Beijing: Zhongyang minzu chubanshe.

Coleman, James S. 1968. "The Concept of Equality of Educational Opportunity," *Harvard Educational Review* 38 (1): 7–22.

Convention against Discrimination in Education adopted by the General Conference of UNESCO in December 1960. Available online at: http://www.unesco.org/education/educprog/50y/brochure/maintrus/28.htm.

Liu, Baocun. 2001. "Kendingxing jihua taolun yu Meiguo shaoshu minzu gaodeng jiaoyu de weilai zouxiang" [Debates on the Affirmative Action Program and Future Trends for Minorities in American Higher Education], *Xibei minzu yanjiu* 30 (3): 172–173.

Ma, Hemin, and Xupin Gao. 1998. *Jiaoyu shehuixue yanjiu* [Educational Sociology Research]. Shanghai: Shanghai jiaoyu chubanshe.

Ministry of Education (MOE). 1995. "Shengshizizhiqu shaoshu minzu jiaoyu gongzuo wenjian xuanbian (1997–1990)" [Selected Provincial, Municipal and Autonomous Regional Documents on Minority Education, 1997–1990]. Chengdu: Sichuan minzu chubanshe.

Song, Taicheng. 2002. "Minzu yuke jiaoyu jianshu" [Summary of Minority Preparatory or Remedial Education], *Minzu Jiaoyu Yanjiu* 53 (4): 16–18.

State Ethnic Affairs Committee (SEAC). 1996. *Guojia minwei wenjian xuanbian (1985–1995)* [Collection of Policy Documents of the State Ethnic Affairs Committee, 1985–1995]. Beijing: Zhongguo minhang chubanshe.

Teng, Xing. 2004. "Xiaokan shehui yu xibu bianyuan pingkun diqu shaoshu minzu jichu jiaoyu" [Moderately Prosperous Society and Basic Education for

Minorities in Remote and Underdeveloped Regions in Western China], *Yunnan minzu daxue xuebao* 21 (4): 148–150.

Wang, Minggang. 2003. "'Jiji chabie daiyu' yu ' jiaoyu youxian qu' de lilun gouxiang" [Theoretical Consideration of the Concepts of "Positive Preferential Treatment" and "Educational Development Priority District"], in *Minzu zhengce yanjiu wencong, dierji* [Studies of Minority Policies, vol. 2], ed. Tiemuer and Wangqing Liu, 285–294. Beijing: Minzu chubanshe.

Wang, Tiezhi. 2001. *Xinshiqi Minzy Zhengce de Lilun yu Shiqian* [Theory and Practice of Minority Policies in the New Era]. Beijing: Minzu chubanshe.

Wen, Jing, Gongning Mao, and Tiezhi Wang. 2001. *Tuanjie jinbu de weida qizhi—zhongguo gongchandang 80 nian minzu gongzuo lishi hugui* [Great Banner of Unit and Improvement—80 Years History Review on Ethnic Work since the C.C.P founded]. Beijing: Minzu chubanshe.

Xinshiqi fazhan minzu jiaoyu de teshu cuoshi [Special Measures for Developing Minority Education in the New Era]. Available online at: http://www.e56.com.cn/minzu/Nation_Policy/Policy_detail.asp?Nation_Policy_ID=448 (July 24, 2005).

Xu, Qingyu. 2000. "Shilun jiaoyu yu jiiaoyu fenliu de guanxi" [On Education Equality and Different Education for Different Students], *Huadong shifan daxue xuebao* (educational science edition) 18 (3): 25–26.

Zhao, Qingdian. 2002. "Dui gaodeng xuexiao zhaosheng tizhi gaige de sikao" [Thoughts on the Reform for College Admission], *Zhongguo gaojiao yanjiu* 1: 56–57.

Zhaosheng kaoshi zhong dui shaoshu minzu kaosheng shixing de teshu zhengce shi ruhe fazhan yanbian de [How Have Preferential Policies for College Admission for Minorities Evolved?]. http://www.e56.com.cn/minzu/nation_policy/Policy_detail.asp?Nation_Policy_ID=288 (July 24, 2005).

Zhou, Yong. 2002. *"Shaoshuren quanli de fali"* [Jurisprudence on Minority Rights in International Law]. Beijing: Shehui kexue wenxian chubanshe.

Chapter 5

Yunnan's Preferential Policies in Minority Education since the 1980s: Retrospect and Prospects

Yanchun Dai and Changjiang Xu

Yunnan Province, located on China's southwestern frontier, has distinctive geographical characteristics and a unique multiethnic history. The cultures of its many minority groups co-exist and blend and are often acknowledged as rich and colorful. However, when China's standardized education was extended to Yunnan in the 1980s, the province's linguistic and cultural differences, as well as factors related to geographic location and poverty, made it difficult for Yunnan minority students to come up to the level of students in China's interior. Thus, the purpose of the province's preferential policies has been to promote educational equality, enabling minority students and those from poor mountainous areas to receive more educational opportunity.

In this chapter, we first review some of the measures taken by the province to improve minority education beginning in the 1980s, and then turn to an analysis of the reasons for problems with some of these efforts as they play out in minority communities and in the context of changing national policy. For example, we take up a point made in the Introduction about the timing of recent preferential policies in minority education in relation to China's transition from a planned, socialist economy to a market economy in the early 1980s. We find a contradiction between China's increasingly competitive market economy, which has created new economic divisions in society, and preferential policies for minorities with their egalitarian agenda. Finally, we conclude our chapter with a set of recommendations to address some of the current problems in Yunnan's education programs for minorities.

The Implementation and Achievement of Preferential Policies

While the programs reflecting preferential policies for minorities over the last three decades are varied in purpose, timing, and locale, they can be succinctly characterized with the following generalizations. For the first nine years of compulsory education, called "basic education," the government has established full-time and part-time boarding schools, introduced policies to supplement or eliminate school fees, and implemented bilingual education. At the middle school, high school, and other secondary levels, preparatory classes have been organized to help students prepare for school entrance exams. Furthermore, when minority students are ready to begin their studies, they are identified for special consideration and encouraged to enroll. There are also preferential policies for teachers from minority areas and remote regions.

Preferential Policies in Basic Education

The main preferential policies in basic education for Yunnan minority students in poor mountain areas include three measures. The first is the government's support for full-time and part-time boarding schools, a measure that concentrates limited educational resources and also guarantees basic living conditions for students. The second is a waiver of school fees and provision of monetary aid in order to lighten the financial burden on students' families. The third is a provision for the development of bilingual materials and bilingual teacher education so that minority students can study Chinese and attain basic knowledge in the standard curriculum while preserving their native languages. Many of Yunnan's minority students live in mountainous areas where transportation is undeveloped, and the population sparse. Because in many areas primary school children must walk one or two hours to get to school, along the way climbing mountains and crossing streams, it is difficult for them to find the energy to study, not to mention dangerous for them in bad weather. Therefore, the provincial government, beginning in 1980, devoted special funds to building a large number of boarding schools. This effort continues to this day.

In addition to the education funding from the central government, since 1980 Yunnan has annually allocated local revenues of 5,500,000 RMB (US$3,294,000 in the 1980s and now just about US$800,000) specifically for establishing boarding schools (Yunnan 2004, 188–301).[1] In 1980, the provincial government also selected for renovation and reconstruction a number of well-managed middle and primary schools in border or mountainous areas inhabited by minorities. A total of nineteen middle schools and twenty-one primary schools benefited from this program. These schools enrolled minority students with financial difficulties and gave each student a monthly stipend of 15 RMB (about US$9) for room and board.[2] By 1981, there were sixteen full-time boarding schools at the middle school and beginning high school levels, with a total of 1,371 students. Another eighteen full-time boarding schools enrolled 1,218 primary school students. Subsequently, the goal was for every township in minority or poor mountainous areas to have one major, well-managed part-time boarding school exclusively for primary school students.

By 1985 there were 1,034 full-time boarding schools for minorities in Yunnan and by 1987, 4,148 part-time boarding schools. Minority students who were in upper primary schools (grades 5–6) and lived far from school boarded and received 7 RMB (US$4) each month to help with their living expenses.[3] As a result, minority students who were enrolled in upper primary schools came to make up 34 percent of the province's total enrollment at that level. In addition, in order to solve the problem of student living conditions, the government coordinated cooperation among related provincial departments. For example, in 1988, the government requested that the grain administration permit students to use the grain they brought with them to exchange for the grain (rice) commonly used in schools. Furthermore, the townships and villages set aside land for growing vegetables and for collecting firewood, as far as possible enabling schools to provide meals for students and ensuring the quality of learning and instruction. In 1999, the level of the stipend for students in full-time and part-time boarding schools was raised from 15 RMB and 7 RMB to 25 RMB (about US$4) and 12 RMB (about US$2), respectively.

Starting in 2000, a portion of the students in border areas attending boarding schools benefited from the province's elimination of school fees for books, school supplies, and various miscellaneous charges. Recently, another provincial program in 2005 eliminated additional fees and supplemented the cost-of-living expenses at boarding schools, benefiting 1,200,000 students. By 2006, another government grant for village education supplemented boarding school expenses for 1,118,000 students (150 RMB/US$22 per primary school student and 250 RMB/US$37 per secondary school student). In 2007, stipend levels continued to increase for poor rural students (250 RMB/US$37 per primary school student and 350 RMB/US$51 per secondary school student) in boarding schools. Minority students from small populations and in poor counties received 300–350 RMB(US$44–51) per person. Students in Tibetan communities received 1,000 RMB(US$147) per person. More than 2,000,000 students benefited from the program (Yunnan 2008, 40).

In past years, boarding schools have concentrated each area's best educational resources for training large numbers of minority student graduates of primary and middle schools. According to statistics from 1986, the rate of promotions in nineteen of the province's minority boarding schools at the primary school level exceeded that for all the "complete" primary schools (those with six grades divided into junior and senior sections) in the same areas. Furthermore, the proportion of minority students in Yunnan's primary and middle schools, relative to the total number of all provincial students at these levels, went up: the proportion of minority middle school students increased from 16.9 percent in 1978 to 26 in 1989. In this same period, the proportion of Yunnan's primary school students who were minorities rose from 27.3 percent to 33.2 (Yunnan 1995, 741).

Overall, the statistics showing the positive results of school consolidation for minority students are remarkable. As a result of the government's consolidation and strengthening of boarding schools, by 2006 there were forty-one province-run boarding schools at the primary and middle school levels. Of these, five were provincial level "complete" middle schools (having both junior and senior high schools). Part-time boarding schools at the primary school level had grown from the original three thousand to fifty-five hundred (Yunnan 2008, 40).

As can be seen from the previous discussion of boarding schools, in the late 1980s the government began to provide supplemental allocations to help many minority students with cost-of-living expenses. In some cases, minority students were exempted from school fees in order to ensure the basic conditions for living and studying typical of poor minority areas. Overall, the government's investment in education has gradually increased. The government has kept its eye on a step-by-step expansion and paid more attention to differential treatment for different students.

As early as 1950, Yunnan's education department mandated that the proportion of students at all levels entitled to the waiver of fees for registration, textbooks, and supplies ranged from 25 to 45 percent. Minority students were given priority under this mandate. This policy has been in effect ever since. Afterward, the practice of eliminating fees or providing financial supplements was mainly concentrated in schools run by the province. The schools established by the province along the border enjoyed a larger percentage of fee waivers. In 1972, it was decided that minority students living within twenty kilometers of the province's borders with other countries should benefit from the elimination of fees and financial help. In 1980, after the interruption of the Cultural Revolution (1966–76), scholarships for secondary school students were restored. Students living in minority border areas were exempt from study fees, book fees, and mimeograph fees, but not all fees. In 1984, the government first eliminated miscellaneous school fees for primary students in the lower grades in mountainous areas. By 1999, students living at the nation's boundaries and attending village schools had no school fees at all.

In 2000, when Yunnan began another round of eliminating some school fees for village-run primary schools, students no longer had to pay fees for notebooks, writing materials, and miscellaneous expenses. In 2004, this policy was extended to all border township-run primary and middle schools, to such schools in minority communities with small populations, and to such schools in Tibetan communities. Up to 2005, this policy for the elimination of fees totaled an investment of more than 200 million RMB (about US$29 million) benefiting 1,280,000 students. Beginning in 2005, when the policy eliminating some school fees was linked with the one for supplementing the cost of living, all of the students enrolled in the compulsory education programs were no longer required to pay miscellaneous fees, and 2,500,000 students enjoyed exemption from fees for notebooks. This represented a further expansion of the policy to reduce the amount of school fees.

The provincial government's policy for support of bilingual education can be seen mainly in the laws ensuring minority language rights in schools, the provision of government financing for bilingual teacher training, government sponsorship of bilingual education reforms and experiments and promotion of successful experiences, and organizing the means for the translation and publication of teaching materials in minority languages used in compulsory education.

In the 1980s, under the auspices of the provincial education department, Dehong and other areas in Yunnan carried out an experiment in bilingual education aimed at finding the best methods for raising the level of minority students' proficiency in both their native languages and Chinese. Their experience was then extended to other areas. In 1988, the provincial education department mandated that when primary school students studying a minority language and Chinese took their promotion exam, their score in their native language could count (up to 30 percent) toward the total of the language score. In general, in prefectures where minority languages were spoken, as well as in their related offices, the minority language score could be one of the criteria for recruiting students, hiring cadres, and reviewing job performance.[4]

By 2006, there were 707 primary schools that implemented bilingual education under the auspices of the provincial education office. This program affected 58,300 students, or about 39 percent of the total number of students in the province who were taking bilingual classes. The province prepared and published teaching materials for bilingual education, including 1,000,000 copies of 276 volumes for 14 minorities in 18 languages. Beginning in 2007, in addition to the series of publications of bilingual materials especially for grade 4, the province began work on a project for translating Chinese textbooks for grade 1 into Dehong Dai script, Xishuangbanna Dai script, Yunnan Miao script, and Sichuan Miao script (Yunnan 2008, 41; for the different scripts of the Dai and Miao languages, see Zhou 2003).

Preferential Policies in Secondary and Higher Education

The most important preferential policies for secondary schools and colleges include the following: lowering the required admission test scores to give priority to minority students; setting up preparatory classes and special prep schools (affiliated with universities) in order to prepare a talent pool of minority students for college education; and establishing a stable corps of qualified teachers.

The principal preferential policy for recruiting minority students is to lower the required score for admission to college and technical schools based on the level of economic development of a particular minority area, and then give these students priority in admission. The most recent preferential policy has been implemented since 2000. This policy uses both ethnicity and level of socioeconomic development as criteria in preferential college admission. (This policy has answered some of the questions raised in chapter four.) First, for border counties, which are usually underdeveloped, minority college applicants receive thirty bonus points on their college admission examination, and Han applicants who were born there or have lived there for ten or more years receive twenty bonus points. This gives minority students a slight edge over Han students from the same area. However, these students' bonus points will be reduced by ten if they go to high schools in the interior. Both minority and Han students are deemed to be somewhat disadvantaged vis-à-vis students from China's interior because of the lack of educational resources and lower education standards for all students in border counties. Second, Yunnan's minority college applicants coming from China's interior receive ten bonus points on their college admission examinations, but Bai, Hui, urban Yi, urban Zhuang, and Manchu applicants do not receive any bonus points. Third, for underdeveloped counties in high mountainous areas, minority students receive twenty bonus points on college admission examinations. Fourth, minority applicants from the Bai, Hui, Naxi, Mancu, urban Yi, urban Zhuang groups, as well as those from groups outside Yunnan, are given priority over Han applicants in admission when they have the same college admission score as Han applicants have. At the same time, the methods for enrolling students in adult education, technical secondary schools, and vocational schools are similar to those described earlier.

In 2003, to attract college students for entry-level government and teaching jobs in border areas, a policy was created to add five points to the college admission examination scores of all college applicants who commit to work after graduation in twenty-five counties along the province's borders and three counties in Diqing prefecture. Excluding Bai, Hui, Naxi, Mongolian, and Manchu applicants, all other minority applicants who have made this commitment receive an additional five bonus points on their college admission examinations or an additional ten bonus points for members of any ethnic group with a population smaller than one hundred thousand.

Preparatory classes are another type of preferential program for minority students. When preparatory classes, including those affiliated with universities, were first offered, their purpose was to serve minority students and those from poor areas by increasing their opportunities to receive higher education and technical secondary education. Preparatory classes at the Yunnan Nationalities Institute began to recruit students in 1980. The program uses a quota from secondary normal schools to admit minority middle school graduates and gives them a corresponding financial packet. They participate in the preparatory classes for three years, and then take the college admission examination. Those who do not get good enough scores to go to college return to their native areas to serve as primary school teachers. Now some of Yunnan's universities and technical secondary schools have also opened preparatory classes, admitting minority students and students from poor areas.

Since 2000, the government has implemented a special policy for relatively less advanced minorities, including the Hani, Miao, Lisu, Lahu, Wa, Yao, Jingpo, Tibetan, Bulang, Buyi, Achang, Nu, Jinuo, Demao, and Dulong. If in any given year there are no high school students from these groups reaching the cutoff points for college admission, then nonetheless two or three of the best among them are selected to attend preparatory classes.

Also starting in 2000, preparatory classes have opened two-year college and high school programs for students from minority groups with relatively small populations. In addition, some college preparatory classes have designated special cutoff points for the admission of minority students from numerically small ethnic groups. In 2007 alone, nearly six hundred students from these groups continued their education in these preparatory classes.

Preferential Policies for Teachers

Yunnan's preferential policies for teachers of minority education include special programs to recruit minority education trainees, special stipends and awards, special considerations for the educational needs of teachers' children, retirement benefits, and term teaching duties in schools in underdeveloped areas.

Yunnan's special programs recruit teacher trainees for minority education in designated areas. Unfortunately, the province's funds for training teachers in basic education do not satisfy the demand, in terms of either quantity or quality. However, owing to the special needs of bilingual education and the harsh living conditions in the province's poor areas, it is entirely feasible to recruit a portion of these teachers from among the minorities themselves. Thus, the teacher trainee programs are directed toward recruitment among minority students, with the proviso that after they complete their studies, they will return home to teach.

Yunnan Normal University and each prefecture's two–three year teachers' colleges have been recruiting students since 1983. Each year they send 3,000 teachers back to minority areas. By 1985, Yunnan Province had a total of 66,466 minority people staffing their education system, among them 57,796 teachers. From 1979 to 1987, Yunnan produced 1,100 minority teachers (Yunnan 1995, 715). In 1986, the allocation system for recruiting and graduating students at the teachers' colleges was reformed. Secondary normal schools were designated to recruit trainees from specific minority communities, students who then return to their home communities after graduation. In 1987, the system for recruiting students was made explicit, thus gradually achieving the "localization" of primary and lower middle school education, where minority teachers are employed in their home communities. In addition, some places began to identify children in rural villages to prepare them for entering minority primary, middle, and normal schools. After graduating, these students returned to their villages to teach, thereby resolving the difficulty of selecting and appointing qualified teachers.

Since 1988, the provincial department of education has taken the following steps to deal with the problem of recruiting teachers. Recruitment quotas for minority teachers' colleges are assigned to counties. Each county then allocates the quotas appropriately among the "hardship" townships where there is a serious lack of qualified teachers. Minority students from the hardship townships applying for college under this quota system can have extra bonus points added to their college admission examination, but their foundation in language and math has to be good. If no "hardship" students come out of this process, then applicants from other townships may be recruited. After their graduation, these applicants have to teach in schools in the hardship townships for five–seven years before moving on.

A second measure to recruit and retain teachers for minority schools is the use of special stipends. In 1982, throughout the province, teachers working in mountainous areas or minority areas received top level stipends for border region intellectuals. To create a stable corps of teachers, each area increased the government's financial benefits for teachers in minority areas, in accordance with local circumstances. For example, Chuxiong Prefecture decided to allocate to normal school graduates an extra step of salary once they report for teaching. It also encouraged young and healthy teachers from urban schools of relatively high quality to take up teaching positions in the mountainous countryside for a five-year term. Each year, they were reimbursed for travel expenses for two round-trips between their schools and homes.

A third measure is the use of rewards, both material and nonmaterial. In 1986, Yunnan decided that education workers with more than twenty-five years in minority areas would be given an honorary designation and preferential material rewards. In 1987, Yunnan allocated five thousand government employment quotas for community teachers not on the government payroll. The provincial government especially requested that priority be given to local minority community teachers for the transfer from community employment to state employment, thus enabling them to receive more stable and higher wages and benefits. Since 1995, the teachers who work in counties identified by the province as "hardship counties" have received an extra salary step and also enjoyed rural teacher stipends, which are provided by prefectural and municipal governments. Other rewards for long-term teachers in minority areas include bonus points for their children's college admission examinations if they apply for teachers' college, and special programs after retirement, including settlement in urban areas, resettlement in their hometowns, or reunification with their children wherever they live in Yunnan. The newest measure taken since 2000 is a six-year term contract, plus extra salary steps for all college graduates who teach in schools in minority and poverty areas in Yunnan.

Summary

All the earlier mentioned preferential policies have worked together to improve minority education in Yunnan. By 2007 there were 2,571,700 minority students enrolled in schools at all levels in Yunnan. They constituted 32 percent of all students, a percentage roughly corresponding to their proportion in the province's total population. In part these numbers reflect the success of preferential policies for minority students in basic education, and in part they reflect the success of preferential policies at higher levels designed to help minority students progress through middle school and beyond. The number of minority students enrolled in primary schools has dramatically increased, and, concomitantly, so has the number of those who are positioned to move higher up on the educational ladder. While Yunnan's total population, based on statistics for 2005, has increased less than 3-fold since 1950, minority enrollments in primary schools have increased nearly 75-fold for the same period, and minority enrollments in middle schools are nearly 1,224 times the number enrolled in 1950, when there were only 649 minority middle school students (He 2005). The teacher corps serving minority areas has also grown. As recently as 2004, statistics revealed that the percentage of teachers in minority and remote areas who are graduates of two-year teachers' schools, three-year normal colleges, and four-year normal colleges, respectively, has reached 95.69 at the primary school level, 93.91 at the lower middle school level, and 80.22 at the high school level.

Some Persistent Problems in Minority Education and their Sources

Minority education in Yunnan has experienced several persistent problems, though great progress has been made. These problems lie in the lack of sufficient funding, student family financial status, student retention, media of instruction, and quality of teachers. These problems may be mainly related to two major factors: the clash between elite education and mass education and the conflict between the state's standardized education and Yunnan's geographic, cultural, and ethnic diversity.

Persistent Problems

Yunnan's investment in education for minorities since the 1950s has grown by leaps and bounds, and, as we have shown, the result has been remarkable growth in the number of minority students in provincial educational institutions and their progress up the education hierarchy. However, the financial resources devoted to minority education and their effectiveness are still inadequate.

The lack of funding may be overt and covert. Overt lack of funding easily attracts our attention. For example, investment in schools has been insufficient to address the problem of unsafe buildings. As recently as 2005, 5.19 percent of the province's school buildings still failed to pass safety inspections. However, covert lack of funding does not often attract attention, and it is more difficult to resolve. Covert funding insufficiency is usually the negative consequence of policy problems. One example is the lack of funding caused by preferential policies that waive all kinds of fees and provide free room and board for minority students, concomitant with the educational funding policy at the national level in the 1990s mandating local management and local funding of primary and secondary schools. When the preferential policies are in effect, a large portion of educational funding is depleted. But the local governments are not able to provide funds to cover the shortage because these governments located in poor areas never collect sufficient tax revenues for public services. The uncoordinated policies lead to an embarrassing difficulty. In poor counties, schools have the best buildings in town, constructed with local and outside funding to meet national school building codes, but they do not have money to pay for electricity, telephone bills, and teachers' supplies (Ma 2000). In addition to facility costs, there is also competition between the teachers' salary pool and the school operating budget. To recruit and retain qualified teachers, schools have to use most of their funds for teachers' salary and benefits, with the result that there is little left from local funds to operate the school (Xin 2001). This is very common in Yunnan: out of Yunnan's 129 counties, about 100 never have sufficient revenues and rely on supplemental provincial and national funding support for 80–93 percent of their budgets (Zhuang and Lai 2002).

This funding shortfall has a negative impact on the implementation of preferential policies. Due to the lack of a system of supervision, the targeted beneficiaries of preferential policies may not benefit from the policies. For example, to keep their ranking and to obtain additional funds, some quality minority schools deliberately downsize provincial student recruitment quotas in order to recruit fee-paying students. The result is not only a reduction in quotas, but a reduction in the actual number of local minority students attending the minority school.

As the burden of supporting education expenses devolved to the local community, local governments and students' families were barely able to support their students. After the financial reforms of the 1990s, the national government and the province in effect withdrew from directly overseeing the administration and implementation of funding for basic education. Although many fees were eliminated through government mandates, expenses were still too much for students' families, so local governments had to assume the burden for education in their respective districts. For example, before the fee elimination program was implemented, Bingzhongluo Township of Gongshan County was supposed to collect as much as 30,000 RMB (US$4,000) in fees, but it actually never collected more than 10,000 RMB (US$1,300). According to the local measures to implement the compulsory education law, the township also was supposed to collect a fine of 50 RMB (US$6–7) from families that failed to send their school-age children to school. But it rarely succeeded in collecting the fine because those minority families were too poor (Zhu 2000). In the end, many of these local measures were unenforceable.

From the point of view of families with students at boarding schools, fee reduction policies still leave many financial responsibilities on their doorstep. And these families, mostly rural and poor, lose the labor that their students formerly invested in family livelihood. Boarding schools in particular entail a per-student cost of living that exceeds that of a student living at home, and all schools get more expensive as the student moves to higher levels and farther away from home. For example, for one Bulang family in Yunnan with two children in school in 2006, one in primary school and one in lower middle school, out-of-pocket expenses for the primary school student ranged from 480 to 600 RMB (US$68–86) per semester, even after fee elimination. For the child in lower middle school, the family's share of education expenses went as high as 2,100 RMB (US$300) per semester. Even with all the support provided by the preferential policies, this was too much a burden for this family, and one semester later the primary school student dropped out of school (Mao 2005, 24). Minority families in poverty-stricken areas face a dilemma. They are worried when their children cannot move up the educational ladder because it means their children will have no socioeconomic mobility. But they are also worried when their children can move up the educational ladder because they cannot afford it. The farther their children climb up the educational ladder, the more financial risk the families take. They will lose all their educational investment if their children fail to pass the college admission examination.

In Yunnan, high levels of enrollment in primary school coexist with low levels of completion. As a result of the national drive for nine-year compulsory education in general and six-year compulsory education in underdeveloped areas, almost all school-age children go to school in minority communities. However, beyond the compulsory years the retention rate has decreased and the dropout rate has increased significantly In some areas, such as in Honghe Prefecture's border counties, dropout levels are up to 13 percent, far beyond the 3 percent set by the national government (Li 2007).

A related problem among primary school dropouts in minority areas is their descent into illiteracy once they leave school. This is because their home environment provides little support for maintaining, let alone improving, their Chinese. We think this situation is one factor in the common perception among minority families that education does not effectively teach their children much, and is thus a poor investment. Bilingual education has not been able to remedy this problem. With little experience in a Chinese-speaking environment and taught by poorly trained bilingual teachers, minority students are often intimidated by standardized national, provincial, and prefectural examinations in Chinese. Moreover, the content of bilingual teaching materials is often remote from the experience of minority students. All these intimidating and frustrating factors contribute to minority students' higher dropout rate.

Despite all the progress in minority education, there is no denying that minority students not only have difficulties in taking the exams but also in re-adjusting to their home environments when they fail the college and technical school admission examinations. This is especially noticeable among the 90 percent of high school graduates who do not qualify to go on to college (Chen 2000). Their long schooling has diminished their identification with their local communities (Jia 2003). They often lack the skills and mindset to find a job; at the same time, they are no longer familiar or satisfied with the rural lives of their parents and peers at home. So they become somewhat like parasites, hanging out in the internet cafes and billiard parlors of the countryside's small towns.

Finding qualified and, more importantly, competent teachers for Yunnan's minority areas is still a problem. For example, in Honghe Prefecture, at the secondary level only 88.5 percent are qualified, with an even lower percentage (75) of qualified teachers in the counties along the border. Even those who are officially qualified are not necessarily competent and knowledgeable. This is a problem largely caused by the preferential policies to localize teacher trainees and teacher training. Teachers who come from the local community, receive their education in the local colleges, and teach in local schools do not have the broad experience and open minds to cope with new challenges. Their limitations constrain the overall development of their students who, in their turn, will become even more limited teachers, thus creating a vicious cycle in basic education (Ma 2000).

Conflicts between an Elite Education System and the Quest for Mass Education

The 1980s mark the creation of the division between two opposite developments in the distribution of educational opportunity in China (Li 2004). Before 1978, the distribution of educational opportunity had evolved into one of the nation's main means for advancing socioeconomic equality. Preferential policies enabled the allocation of opportunities in education and even education models to favor students from workers', peasants', and minority families. Thus, it did not matter whether the planned economy was the mainstay of the economic system or socioeconomic equality was the goal of the education system; both were consistent with the egalitarianism of preferential policies for the education of minorities. From financial support and guarantees to the allocation of employment for graduates in the old planned economy, education for minorities was assured. After 1978, for China to develop its economy and link up to the world, to move away from "poor and backward" on the margins of modernization, and to become "rich and civilized" and at the center of modernization, the purpose of education changed accordingly. Its goals became to advance the country's economic development and achieve modernization by identifying and training a talented elite, and to advance a consumer economy by promoting the mobility of human and social capital across dissimilar areas. In short, education was transformed from a means for ensuring socioeconomic equality into a mechanism for the production of economic divisions (see Yunnan governor's education policy statement; Yunnan 1995, 713). The egalitarianism of the education system came to an abrupt halt while an elite education system has since appeared. For an understanding of minority education policies made and implemented since the 1980s, this background is essential. As the mechanism for distribution devolved from government and the nation to the marketplace, differences among localities have become even more pronounced. Although the central government's preferential policies for minorities promote the distribution of benefits among disparate localities, in the end these policies cannot completely replace the local government's responsibility for the development of disadvantaged areas.

From the point of view of the local government, the responsibility for minority education has fallen to them. Central government funds have dried up, and preferential policies in minority education have resulted in the reduction of educational funds from local families. The local government has the responsibility to oversee education but lacks the funds to do so. As we have seen, from the point of view of local farm families, considering the cost and risk involved in educating their children, abandoning education altogether cannot but have a certain logic. Assuredly, doing so is not due to the "backward thinking" so often attributed to peasants and minorities. If education endangers subsistence, it is no wonder local minority families have reservations. As long as education is tailored for urban students, the targets of minority education will continue to ignore and underrate it.

The Contradiction between the Standardized Education System and Yunnan's Natural and Multicultural Environment

The distribution of today's educational resources has been influenced by the market and is relatively concentrated in the country's urban areas. Of course minorities thinly dispersed over remote areas with poor transportation are likely to fall below the requirements of the standardized education system. Under the circumstances, the education system must adapt to these conditions. This is minority education's biggest challenge.

The ideal distribution of main primary schools and their branches for efficiency and convenience has been an issue in Yunnan for many years (for more, see chapter eight). After many years' adjustment, the current distribution pattern is like this: lower level or branch primary schools (grades 1–3) are located in villages and upper level primary schools (grades 4–6) and middle schools are located in township seats. The advantage of this pattern is the convenience for younger children of attending school, but the disadvantage is the lack of quality educational resources in branch primary schools. For students just beginning primary school, their educational foundation is quite weak, and inferior educational resources negatively influence their subsequent educational careers. From the point of view of efficient use of limited educational funds, given the transportation difficulties and the nature of the land, the costs of funding for building schools, village transportation, excavating the land, leveling the land, and so on is necessarily a large portion of the budget. From the point of view of investment in human resources, we must be aware that teachers in remote mountainous areas have many functions outside the classroom—student guardians, nannies and cooks, for example—and must endure many hardships. These functions have prevented them from attending in-service training and have affected their focus on teaching.

Cultural and linguistic diversity is Yunnan's special feature but also its challenge in developing education. For example, Yunnan's minority languages present many challenges to bilingual education. There are over twenty minority groups, who speak twenty-four officially recognized languages and use sixteen written languages. Developing bilingual materials and training qualified teachers, translating standard textbooks into various minority languages, and training bilingual teachers to use standardized syllabi and textbooks for so many linguistic and ethnic minority groups prove to be difficult (He 2005).

Recommendations

The Yunnan provincial government has always paid attention to and supported the education of minorities. The province has already mandated a series of laws and policies and adopted effective measures for advancing the development of minority education. The preferential policies for minority education discussed attest to this point. However, owing to geographical and historical factors, many difficulties remain. To create a maximally effective program of minority education, the government should consider additional steps to alleviate five difficult issues: the cost burden on families, the poor quality and unsafe condition of schools, the chronic teacher shortage, the linguistic barriers for students, and the continuing imbalance in the allocation of educational resources.

The cost burden on families results from the expense of boarding students away from home. Increasing financial aid so that every primary school student receives 800 RMB (US$120) per school year and every junior high school student 1,000 RMB (US$143) per year would go a long way toward solving this problem.

Financial investment can also solve the second issue: the province needs to increase capital investment in school construction to ensure student and teacher safety. This, in turn, will also help to solve the third issue: Safer schools and improved teaching conditions will help to attract good teachers. In addition, the province can enact more preferential policies for teachers, allow substitute teachers to take a test to become official teachers, provide hardship pay for the neediest areas, and change the teacher rotation policy so that the length of time served is reduced. Perhaps it should implement a policy requiring all new teachers in the province to serve in minority areas for three years. As these policies result in more teachers overall, class sizes can be reduced and teacher in-service training opportunities can be increased, which will in turn continue to attract more teachers.

To address the cultural and linguistic gap between minority students and their Han counterparts, the province should begin to train more bilingual teachers in order to facilitate bilingual education. For example, in areas where the level of Chinese is not high, it should add one year of preschool classes where Chinese study is the main focus. It should also pay special attention to the localization of textbook knowledge to make education more relevant to students and their families. In the last semester of lower middle school, it should add courses on agricultural and applied technology and knowledge.

The final issue can be addressed partly through increased financial investment in teaching materials and supplies, and partly through implementation of long-distance education. By realizing the goals of educational information technology, Yunnan can substantially remedy the imbalance in education resources.

Notes

1. To clarify, all policy documents and data used in this chapter, unless otherwise indicated, are from the publication Yunnan Provincial Ethnic Affairs Committee and Yunnan Ethnic Studies Association (eds.). 2004. *Minority Policy Abstracts, 1949–2003*. Pp. 188–301. Kunming: Yunnan Provincial Ethnic Affairs Committee.
2. The U.S. dollar value for the Chinese renminbi (RMB) varied during the period of 1980s, 1990s, and 2000s. The exchange rate for US$1 ranged from 1.7 RMB in the 1980s to 8.6 RMB in the 1990s, and then to 6.8 RMB in the mid-2000s.
3. There were usually more lower-level primary schools (grades 1–3) scattered in minority communities, but only one or two primary schools with upper levels in a township. That is why upper levels needed boarding schools.
4. For example, after a relatively early experiment in bilingual teaching in Dehong prefecture, in 1978, the prefecture mandated that graduates of primary schools who had taken classes in their own language could take an additional test in their native language, and add that score to comprise up to 30 percent of their total language test score for promotion. Moreover, the tests for recruiting students, workers, and cadres in Dehong also added a minority language test. Primary school teachers of minority languages were given special consideration in determining their salaries and other allowances (see Yunnan 1995, 753).

References

Chen, Xianghong. 2000. "Yunnan minzu jiaoyu fazhan de xianshi wenti sikao" [Thoughts on Pragmatic Issues in the Development of Minority Education in Yunnan], *Xueshu tansuo* 4: 82–85.

He, Fusheng. 2005. "Yunnan minzu jiaoyu fazhan de licheng yu zhuyao tedian" [Development and Characteristics of Minority Education in Yunnan], *Yunnan jiaoyu* 12: 26–29.

Jia, Zhongyi. 2003. "Cong Yunnan renkou jiaoshao minzu de diaocha kan minzu jiaoyu de jige wenti" [Several Issues in Minority Education: Study of Cases of Minority Groups with Small Populations], *Minzu jiaoyu yanjiu* 56 (3): 14–18.

Li, Chunling. 2004. "Shehui zhengzhi bianqian yu jiaoyu jihui bu pingdeng: jiating Beijing ji zhidu yinsu dui jiaoyu huode de yingxiang (1940–2001)" [Political Change and Equality in Educational Opportunity: The Impact of Family Background and Social System on Access to Education], in *Zhongguo Shehui Fencong* [Social Stratification in China], ed. Peiling Li, Qiang Li, and Liping Sun, 393–424. Beijing: Shehui kexue wenxian chubanshe.

Li, Jiyun. 2007. "Bianjian qian fada diqu jichu jiaoyu fazhan xianzhuang fenxi: Yunnan yi Honghe Hanizhu Yizu zizhi zhou wei li" [Analysis of the Current Status of Basic Education in Underdeveloped Areas: A Case Study of Honghe Hani and Yi Autonomous Prefecture, Yunnan], *Nongye Kaogu* 6: 435–437.

Ma, Keli. 2000. "Dui Yunnan shaoshu minzu jichu jiaoyu de dangdai sikao" [Current Thoughts on Minority Education in Yunnan], *Yuxi shifan gaodeng zhuanke xuexiao xuebao* 16 (extra edition): 227–230.

Mao, Yaping. 2005. "Shucun pinkun yu fan pinkun wenti gean yanjiu" [A case study of poverty and anti-poverty in Shucun village], Yunnan University, master's thesis.

Xin, Aihong. 2001. "Yunnan minzu zizhi difang jiaoyu fazhan duice yanjiu" [Studies of Solutions to Education in Minority Autonomous Areas in Yunnan], *Yunnan minzu xueyuan xuebao* (social science edition) 18 (4): 77–80.

Yunnan. 1995. *Yunnansheng zhi: jiaoyu zhi* [History of Yunnan: Educational History]. Kunming: Yunnan renmin chubanshe.

———. 2004. *Minzu zhengce jiyao (1949–2003)* [Abstracts of minority policies in Yunnan, 1949–2003] (Internal circulation only). Kuming: Yunan minzu shiwu weiyuanhui.

———. 2008. *Gongtong tuanjie fendou, gongtong fanrong fazhan: wunian lai de Yunnan minzu gongzuo* [Unite and Strive Together for Common Prosperity and Development: Minority Work in Yunnan in the Past Five Years]. Kunming: Yunnan minzu chubanshe.

Zhou, Minglang. 2003. *Multilingualism in China: The Politics of Writing Reforms for Minority Languages 1949–2002*. Berlin/New York: Mouton de Gruyter.

Zhu, Heshuang. 2000. "Shaoshu minzu chuangtong jiaoyu yu xiandai jiaoyu de jiaoxiang bianzouqu—yi Yunnan Gongshan Xiachalu Dulongzu wei li" [The symphony of minority traditional education and modern education: a case study of the Delung in Xiaochalu of Gongshan County, Yunnan], *Zhongyang minzu daxue xuebao* (social science edition) 6: 91–102.

Zhuang, Wanlu, and Lai Yi. 2002. "Minzu diqu jiaoyu xianzhuang yu duice yanjiu, yi Yunnan, Guizhou, Sichuan sansheng wei li" [Current Status of and Solutions to Education in Minority Areas: Case Study of Yunnan, Guizhou, and Sichuan], *Xinan Minzu Xueyuan Xuebao* (social science edition) 23 (5): 21–33.

Part II

Between State Education and Local Cultures

Chapter 6

Anthropological Field Survey on Basic Education Development among Machu Tibetan Nomads

Gelek

Since China's "Open Up the West Campaign" began in 2000, both the central and local governments have given a great deal of attention to developing and improving basic education in ethnic minority areas in western China. The People's Republic of China (PRC) government assigns high value to this campaign and believes it has an important role to play in the nation's modernization, particularly at the local level. In consequence, basic education has been given high priority, and the central government of China has increased investment in basic education in the ethnic minority areas of western China, including Tibetan cultural areas. For instance, in June 2002, the PRC's Ministry of Education began implementing a project called "The Development of Basic Education in Western China." It has lent a total of over US$10 million at low interest rates to five western provinces, namely, Yunnan, Sichuan, Guangxi, Ningxia, and Gansu, in order to support basic education in these ethnic minority areas.

The project area covers ninety-eight counties in the five provinces, including two where Tibetan culture predominates. They are Kangding County in Sichuan and Machu County in Gansu. The purpose of the project is to improve the quality of basic education and to enable students to complete it. The target groups are those children who live in minority areas where economic conditions are poor. The project gives priority to those in the ethnic minority communities lacking the opportunity to go to school.

I was honored to be invited to take part in the project and to act as a consultant for it.

From February 18 to March 18, 2003, I had the opportunity to undertake anthropological fieldwork in Kangding (*Dar Tse Do*) County in the Ganzi Tibetan Autonomous Prefecture, Sichuan Province, and Machu (*Maqin*) County in Gannan Tibetan Autonomous Prefecture, Gansu Province. My aim was to carry out a survey and social assessment of basic education in the Tibetan areas, especially among the Tibetan pastoral nomads.

During fieldwork I held discussions with officials at different levels, including prefecture, county, and township cadres, as well as village leaders. I also visited middle and primary schools in the counties and townships and three Tibetan villages in two townships. With other researchers, I interviewed teachers, students, and herders, as well as children who had not enrolled in school or had dropped out. We also talked in different places to teachers and pupils, both Tibetan and Han, and we visited Tibetan villages and interviewed herders' households. By employing both qualitative and quantitative methods, we were able to collect firsthand data and make use of prefecture, county, and township documents and statistics. This paper draws on some of this data and focuses in particular on the main factors affecting the development of basic education among the Tibetan nomads (*drog pa*) of Machu County.

Machu County as a Case of Basic Education in Tibetan Communities

It is clear that modern education has been developing rapidly in Tibetan cultural regions over the past half-century, and especially since the reform period began in the early 1980s (see Geng and Wang 1989; W. Zhou 2004; Zhou and Liu 1998). Education, which was nearly nonexistent fifty years ago, has developed into a comprehensive structure that goes from primary school to junior high school, senior high school, and university (see R. Zhou 2002). The system has improved with the transition from private to public and from informal to formal schooling. Furthermore, in the Tibetan Autonomous Region (TAR) the number of high schools (both junior and senior) has increased from virtually nil to 2,537 between 1951 and 1990. This means that an average of about 60 such schools were set up each year. Moreover, in 1990 there were 178,700 students in school, which is sixty times more than there were in 1951 in the TAR. In addition, there are students who have joined the School for Tibetan Students in Inland

China (see chapter seven on *neidi* schools). By 1990 about 80 percent of the total school-age population was attending school in the TAR.

However, if you compare only the basic educational development level of the TAR with other provinces in China, the student dropout rate still remains too high, not only in Machu and Kangding Counties, but also in other Tibetan areas in Sichuan, Qinghai, and Yunnan Provinces (for Tibetan literacy rates and educational attainments in comparison with other ethnic groups, see M. Zhou 2000, 2001, 2007). Here is a specific example of primary school grade progression for Machu County.

Table 6.1 shows the number of male and female students in each grade from grade 1 to 6 during a five-year period from 1999 to 2003 in Oula Township Primary School in Machu County. We can see that about half of the first graders moved up to grade 2, and about a quarter or less of the first graders went on to grade 3. Very few moved beyond grade 3. This situation is found throughout the TAR, where many areas continue to have high dropout rates. Tibetan communities' lower primary school graduation rates affect secondary school enrollment and graduation rates as well, particularly in comparison with the TAR's neighboring provinces/regions. While the number of secondary school students in the TAR almost doubled (to a little over one hundred per ten thousand people) in the 1990s, this region still had the lowest number of high school students per ten thousand people compared with its neighboring Tibetan communities, which reached two hundred–three hundred per ten thousand (for more on Tibetan educational attainment, see M. Zhou 2001). The national census conducted in 2000 shows that only 12 percent of the Tibetans aged six and older had secondary education while nationally 48 percent of this age cohort had secondary education (China 2003: 124–131).

Table 6.1 Number of students from 1999 to 2003 in Oula Township Primary School (T: total; M: male; F: female)

	1999			2000			2001			2002			2003		
Grade	T	M	F	T	M	F	T	M	F	T	M	F	T	M	F
1	45	31	14	58	39	19	70	40	30	79	41	38	70	37	33
2	14	8	6	12	9	3	34	18	16	52	30	22	47	29	18
3	9	5	4	12	7	5	14	11	3	18	12	6	25	15	10
4	3	1	2	10	5	5	12	7	5	13	10	3	12	9	3
5	2	0	2	2	0	2	9	4	5	9	5	4	9	5	4
6	3	3	0	2	0	2	2	0	2	8	4	4	8	4	4
T	76	48	28	98	60	38	141	80	61	179	102	77	171	99	72

Some of the main reasons we found for this phenomenon during our fieldwork included the demand for child labor and the distance between homes and schools in the Tibetan nomadic areas. The attitude of parents toward basic education also has a significant impact on their children's enrollment rates. Different localities have experienced varying degrees of economic development, which also impacts basic education. Before discussing those reasons, I would like to take a little time to introduce basic education in Machu County where I did my fieldwork.

Machu, which in Tibetan means "Yellow River," is located in southwestern Gannan Tibetan Autonomous Prefecture in Gansu Province, just near the borders with Qinghai and Sichuan. It has a population of 38,350, of whom 35,685 are Tibetans, most of them making their living through animal husbandry. Traditionally, the population of Machu included nine groups (*tsho wa*) of Tibetan pastoral nomads: the *Chuka Ma*, *Tsherma*, *sMad-ma*, *Wa Bantshang*, *rMara*, *Zhichung*, *dNgul-la*, *Nyin-Ma*, and *dMer-sKor*.

Nowadays, there are eight townships and fourteen schools in the county. Apart from two primary schools in the county town, all are boarding schools. The schools are as follows:

- one complete primary boarding school in each township, with one township primary school located in the county seat
- two primary schools in two very remote villages of two townships
- one complete primary school for Tibetans and other ethnic groups in the county seat
- two complete middle schools, one for Tibetans only and the other for both Tibetans and other ethnic groups in the county seat
- one nursery school in the county seat

There are two kinds of boarding schools in all Tibetan areas. The first is called *sanbao*, which means "three responsibilities." The government provides food and accommodation, textbooks, and tuition for free. This policy has been implemented in the TAR since 1985, in order to ensure nine-year compulsory education (for the policy document, see Tibet 1999, 424–427). In the second kind of boarding school, the government provides only accommodation and tuition for free. This kind of school is found in the Tibetan areas of Yunnan, Sichuan, Gansu, and Qinghai, including Machu County. All schools in the TAR and the four provinces listed use a Tibetan-language textbook, which was produced in 1992 and has not been revised since then (for more on Tibetan language education, see Maocao Zhou 2004; Zhou and Gesang 2004).

Since the implementation of the "Open Up the West Campaign," the central and local governments have paid more attention to basic education development in Tibetan areas. The Machu County government has taken various measures to improve basic education in recent years and has advanced the specific basic educational principle espoused in the slogan "give first place to boarding schools that are public, full-time and teach in Tibetan." These policies have increased the students' enrollment rate and decreased the dropout rate significantly, not only in Machu but also in other Tibetan areas in Sichuan, as well as in Qinghai, Yunnan, and the TAR. However, the dropout rate is still high.

Main Factors Affecting the Dropout Rate

There are a number of reasons why children drop out from school. The following are the six main factors that have most negatively affected basic education.

The first factor is demand for child labor (see chapter nine). Grazing is one kind of labor suitable for children. The introduction of the household pasture contract system in Machu in the early 1980s not only changed the production mode but also the herders' way of life in the area. Every household contracted two areas of grassland (one for winter and one for summer). Herders settle near the winter grassland in winter and in summer live in tents near the summer grassland. The location of the households is scattered, and they have to divide labor among the family members to take care of their cows and sheep, as well as to take on sideline forms of household production, such as collecting *yartsa gunbun* (caterpillar fungus—an expensive herbal medicine), doing manual work in the county seat, or undertaking other business. In addition, there is a customary division of labor in Tibetan herding areas. Women perform about 60 percent of work in the pastoral areas, which greatly affects access to schooling for girls.

The second factor is parental attitudes to basic education (also see chapter nine). First, most parents demand a return on their investment in their children's basic education. They expect that those children who go to school will get a good job. If this does not happen, parents are likely to lose interest in sending their children to school. When we asked students' parents: "What kind of occupation or job would your children like to take up when they graduate from school?" most answered that they would like their children to become cadres (government officials) or doctors. This view comes from the period of the planned economy, when the government created the expectation that minority students who went to university, secondary specialized school, or secondary technical school would have a guarantee of a permanent job. This system was called the "iron rice bowl." At that time, most Tibetan students got employment as cadres. However, now that the iron rice bowl has been broken, there is no total guarantee that all students will get jobs after graduation from university. There is now a job market, in which talented and resourceful graduates are likely to do much better than others (for more on the job market for college graduates, see chapter ten). Given this less certain kind of result for their educational investments, parents are less likely to want to send their children to school.

The third factor is the distance from school, affecting children's access to basic education (also see chapters five and eight). Machu County is about 10,190 square kilometers in area. In 1990 there were only about three persons per square kilometer. Also, there is only one primary school per township and one Tibetan-language secondary school in the whole county. For many students the nearest primary school is more than 50 kilometers away from their home and for some students, such as those of Muxihe Township, the secondary school is more than 130 kilometers away from home. The distance factor is one of the reasons why boarding schools have been set up in the Tibetan nomad areas. It also accounts for the fact that many children are still in primary school when they are fourteen or fifteen years of age. Their age makes them feel inferior and at a disadvantage, which can also lead to dropping out.

The fourth factor is the inadequate conditions in the students' dormitories in school (for more, see chapter five). In the townships in Machu, all primary schools are boarding schools. Children have to live at school because their homes are so far away. But in many of these schools, such as the Tibetan-language Secondary School in Machu County, living conditions are inadequate. Now that more and more children go on to the higher grades of school, the government has allocated more funds to cope with the increasing demand for accommodation at the schools. However, it is still grossly inadequate. Many parents even rent a house and live close to the school, in order to be able to look after their children while they are at school.

The fifth factor is the inappropriate teaching materials (see Maocao Zhou 2004). The only textbook used was produced especially for the Tibetan communities in the TAR and four provinces. But it is actually mostly a translation from Chinese, and the context is inappropriate to the culture and way of life of Tibetan herders (for the tradition of Tibetan culture, see Ciwang 2001; Gelek 2006; for the modernization of the tradition, see Gelek 2005). For instance, the textbook talks about traffic lights, but children have never seen traffic lights before and do not know what they are. This cultural disconnect makes many students lose interest in their studies. Most teachers we interviewed told us that they lacked teaching materials, such as suitable books, television and video equipment and programs, and so on.

The last main factor is the impact of economic development (see W. Zhou 2004, 324–357). Different localities have experienced varying degrees of economic development, which affects local basic education. For instance, in Guzha village in the Ganzi Tibetan Autonomous Prefecture, there are 384 Tibetans and 75 Han. This village is located on the main Sichuan-Tibet highway, as a result of which trade has developed rapidly there. Most of the villagers run stores, and 60 percent of families own trucks and phones. The average annual income per person is 2,000 RMB. Some families even make over 10,000 RMB per year. Parents there do not need child labor. They also realize that education helps their business. Thus they are very keen to send their children to school, and the enrollment in local primary and secondary schools is 100 percent.

Conclusion

Since the "Open Up the West Campaign" began in 2000, the central and local governments of China have paid more attention to basic education and have taken various measures to improve it in the Tibetan areas. As a result, basic education in the Tibetan cultural areas has developed rapidly in recent years. Although the dropout rate has fallen, it still remains too high, not only in Machu, but also in other Tibetan areas in Sichuan, Qinghai, and Yunnan, as well as in the TAR. Some of the main reasons we found for this phenomenon during our fieldwork in Machu included the demand for child labor and the distance between homes and schools in the Tibetan nomadic areas. The attitude of parents toward basic education also has a significant impact on their children's enrollment rates. Different localities have experienced varying degrees of economic development, which also impacts basic education. This year we were pleased to hear that the PRC's central government has made an important decision to waive tuition for basic education for all students in rural areas of China. I believe this new education policy will be of great benefit, encouraging more Tibetan students to enter school and reducing the dropout rate among Tibetan nomadic students. However, I also recommend the following:

1. Add new courses in Tibetan on local grazing knowledge.
2. Promote changing production and concentrate population density in herding areas.
3. Promote the participation of parents in basic education programs and change their views and attitudes about basic education.
4. Improve student dormitories at township boarding schools.
5. Ensure continuous government support for basic education, such as exempting tuition fees and subsidizing boarding fees.

6. Give teachers language and language pedagogy training. Tibetan teachers should learn Chinese, and Chinese teachers in Tibetan cultural areas should learn Tibetan.

References

China. 2003. *Tabulation on Nationalities of 2000 Population Census of China*. Beijing: The Ethnic Publishing House.

Ciwang, Junmei, et al. (ed.). 2001. *Xizang zongjiao yu shehui fazhan guanxi yanjiu* [A Study of the Relationship between Religion and Social Development in Tibet]. Lhasa: Xizang renmin chubanshe.

Gelek. 2005. *Yueliangxichen de difang* [The Place Where the Moon Sets]. Chengdu: Sichuan minzu chubanshe.

———. 2006. *Xizang zaoqi lishi yu wenhua* [Early Tibetan History and Culture]. Beijing: Shangwu yinshuguan.

Geng, Jinsheng, and Xihong Wang (ed.). 1989. *Xizang jiaoyu yanjiu* [Studies of Education in Tibet]. Beijing: Zhongyang minzu daxue chubanshe.

Tibet. 1999. *Xizang zizhiqu jiaoyu falufagui xuanbian* [Selected Educational Laws and Regulations in Tibet]. Lhasa: Xizang renmin chubanshe.

Zhou, Maocao. 2004. "The Use and Development of Tibetan in China," in *Language Policy in the People's Republic of China: Theory and Practice Since 1949*, ed. Minglang Zhou and Hongkai Sun, 221–238. Boston: Kluwer Academic Publishers.

Zhou, Minglang. 2000. "Language Policy and Illiteracy in Ethnic Minority Communities in China," *Journal of Multilingual and Multicultural Development* 21 (2): 129–148.

———. 2001. "The Politics of Bilingual Education and Educational Levels in Ethnic Minority Communities in China," *International Journal of Bilingual Education and Bilingualism* 4 (2): 125–149.

———. 2007. "Legislating Literacy for Linguistic and Ethnic Minorities in Contemporary China," in *Language Planning and Policy: Issues in Language Planning and Literacy*, ed. Anthony J. Liddicoat, 102–121. Clevedon: Multilingual Matters.

Zhou, Runnian. 2002. *Xizang jiaoyu wushinian* [Fifty Years' Education in Tibet]. Lanzhou: Gansu jiaoyu chubanshe.

Zhou, Runnian, and Hongji Liu. 1998. *Zhongguo zangzu siyuan jiaoyu* [Monastery Education in Tibet]. Lanzhou: Gansu jiaoyu chubanshe.

Zhou, Wei (ed.). 2004. *Ershiyi shiji xizang shehui fazhan luntan* [Twenty-first Century Forum on Social Development in Tibet]. Beijing: Zhongguo zangxue chubanshe.

Zhou, Wei, and Jiangcun Gesang (eds.). 2004. *Xizang de Zangyuwen gongzuo* [Tibet's Work on Tibetan Language]. Beijing: Zhongguo zangxue chubanshe.

Chapter 7

Tibetan Student Perspectives on *Neidi* Schools

Gerard A. Postiglione, Ben Jiao, and Ngawang Tsering

A preferential education policy specifically targeted at the Tibetan Autonomous Region (TAR) is widely considered to be a great success after twenty years of implementation. This policy established what has come to be known as the *neidi Xizang ban* (inland Tibetan schools and classes; Postiglione et al. 2004). This chapter focuses on the part of that policy that sends the top graduates of Tibet's primary schools to boarding schools in China's urban areas for study of up to seven years.[1] Our aim is to review the main aspects of the policy and present preliminary data about student perspectives on their experience at *neidi* schools. In our conclusions, we point out the largely positive benefits of *neidi* schools as well as the concerns of both graduates and the state. The schools and their curricula, in preparing young Tibetans for careers as cadres in the TAR, inculcate in students a Chinese national identity, without sacrificing the students' sense of themselves as Tibetans.

Ethnic Minorities within the People's Republic of China

In the late 1970s, the Chinese state moved toward a policy of economic reform and opening to the outside world. According to Mackerras (1999), this led to a resurgence of ethnicity and greater recognition of ethnic-minority culture that had been subverted to the fanatical focus on class struggle during the Cultural Revolution (1966–76). The designated ethnic-autonomous regions cover half the country and the leadership in these regions provides preferential treatment policies for ethnic minorities in employment, family planning, and education (Sautman 1999). Their laws permit ethnic-minority regional schools to use native languages as the medium of instruction (see Zhou 2003; Zhou and Sun 2004). However, with the spread of the market economy, there is a strong pull toward mainstream Chinese-language instruction, which household heads perceive as opening broader pathways to university study and employment (Ma 2007; also see chapter ten in this volume). As in other countries, China is engaged in debates about native languages, cultural preservation, and economic development. Meanwhile, it continues to subscribe to the notion made popular by Fei Xiaotung of *duoyuan yiti geju* or "plurality and unity within the organic configuration of the Chinese nation" (Fei 1986; Gladney 1995). Educational policies that give preferential treatment are operationalized within this framework, and one of the best known is the practice of sending ethnic minority children to state boarding schools.

Background on Education in Tibet

Tibet is ethnically homogeneous, intensely religious, and geographically remote from Beijing. It is also the poorest major region in the country. Before 1951, it had a high degree of autonomy (Goldstein 1989, 1997). Since then schooling has come to be an increasingly key agent of cultural transmission but access remains a problem for many reasons, including Tibet's size, remoteness, and population dispersion. Following the so-called peaceful liberation in 1950, when the Chinese military entered Tibet and formally declared it a part of the People's Republic of China (PRC), few changes occurred in the traditional theocratic structure of government, the organization of monasteries, and traditional forms of landholding. However, political difficulties led to the Dalai Lama's flight to India in 1958, where he still remains. The TAR was officially established in 1965. It covers 1.2 million square kilometers, 12.5 percent of the area of China, though it is home to only 0.002 percent of the population (Irendale et al. 2001; Zhang 1997). It is located a great distance from mainstream China and its residents possess a distinctive culture dating back over a thousand years, with a complex religious tradition and writing system. Its people live at extraordinarily high altitudes, predominantly plateau, averaging 3,600 meters above sea level and surrounded by mountains. Tibetans are dispersed across a region that stretches far beyond the TAR. More Tibetans live outside of the TAR in the surrounding provinces of Sichuan, Qinghai, Gansu, and Yunnan than in the TAR. In all, Tibetans occupy an area about as large as the continental United States.

Monastery education dominated before 1951, when the Chinese established a school in the capital Lhasa, and when the Seventeen Point Agreement was signed. According to this agreement, "the spoken and written language and the school education of the Tibetan nationality shall be developed step by step in accordance with the actual conditions of Tibet" (Goldstein 1989, 767). By 1959, shortly after the Dalai Lama fled to India, Tibet's educational system was brought closer in line with the rest of China. Nevertheless, monastery education, although tightly controlled, still exerts a strong influence on modern education (Mackerras 1999). It is virtually impossible to separate Tibetan religion from other aspects of Tibetan culture, and the wish of Tibetan families to have one of their children enter the monastery is common in many places. Until the 1990s, monks were often the most literate members of rural and nomadic communities. The cleavage between cultural transmission of the monastery and school is not surprising within the Chinese system (Nyima 2000).

Tibet lags behind China's other four ethnic autonomous regions in the establishment of schools. After the Lhasa Uprising in 1958, education was put on the fast track by the Beijing government, though it still remained a decade behind in matters such as collectivization. The Cultural Revolution tore into the fabric of Tibetan life with devastating results. Class struggle became the order of the day and the quality of teaching and learning, already low, became even worse. Where they remained open, schools became predominantly an ideological arena for propaganda and self-criticism. Class warfare took precedence over academic affairs and any mention of cultural heritage became associated with feudalism and severely criticized. Nevertheless, by the later part of the Cultural Revolution in Tibet, there was an expansion of school numbers. Figures in Bass (1998, 39) show rapid growth in primary school enrollments from 1965, the year the TAR was established, and a leveling off in 1978, when the emphasis shifted from quantity to quality and enrollments began a drastic decline.

With the dissolution of the communes in 1984, many parents withdrew their children from school to labor in the new household economy system. The more open policy after the Cultural Revolution initially led more children to attend monasteries instead of the poorly staffed schools that lacked trained teachers (Bass 1998). The effect of decentralization was to leave rural schools with fewer resources for school buildings, instructional materials, teacher salaries, and especially the reform and localization of school curriculum. Throughout the years, Tibet has developed more slowly than other parts of China in education. Most notably, there has been a lack of qualified teachers.

Tibet *Neidi* Schools: Background

The idea of educating China's non-Han nationalities in schools closer to the cultural and political center of the country is not new. During the Qing Dynasty (1644–1911) and Republican Era (1912–49), schools for Mongols and Tibetans operated in Beijing (also see chapter two). When the *neidi* schools were established thirty-five years after the 1949 revolution, there was no official statement linking them to earlier schools for non-Han Chinese in Beijing, either due to the Chinese Communist Party's (CCP) unwillingness to link current policies to past regimes or because they do not perceive such a connection. Nevertheless, it is generally accepted that *neidi* schools are part of a long-term strategy to build national unity, as well as to train specialized talent for the TAR's development.

The story of the *neidi* schools begins with the visit of Hu Yaobang (CCP's chairman and general secretary, 1981–89) to Tibet after the end of the Cultural Revolution, who found the situation in Tibet extremely unsatisfactory (see Zhu 2007). Urgent discussions took place about how to improve the living standards of Tibetans. As early as April 1980, the central government had called for Tibet to take measures to improve education, including training of specialized talent. Education quality was discussed four years later at a work meeting about Tibet's development. Hu Qili and Tian Jiyun (both vice premiers and members of the CCP Central Secretariat then) called for inland cities to develop talent by establishing schools and classes for Tibetan graduates of primary schools. Most students would be educated in junior-secondary schools as preparation for senior, secondary-level, specialized (*zhongzhuan*) schools, and a small number would attend regular senior-secondary schools to prepare for college or university. Three Chinese cities established *neidi* schools in 1985. Soon after, they were established in sixteen other provinces and municipalities. Financial responsibility was shared by the Tibetan government and host cities (Xizang 2005).[2] Between 1985 and 2005, over 25,000 Tibetan students were sent to study in *neidi* schools across China (N.A. 2007).

The most noted *neidi* schools were the ones in Beijing and Changzhou (Jiangsu Province), the latter founded in 1987.[3] The Beijing school recruited junior and senior students, and the Changzhou school recruited only junior-secondary students. More schools were opened during the 1990s. Nantong Tibetan School in Jiangsu Province was founded in 1995 and accepts both junior- and senior-secondary students. Chengdu Tibet School in Sichuan Province was opened in 1998 for junior- and senior-secondary students, but will drop its junior secondary segment. Chongqing Tibet School, also in Sichuan Province, opened in 1995 and has both Han Chinese and Tibetan students, in a segregated educational format, at both junior- and senior-secondary levels. *Neidi* schools that admit senior-secondary students also include those in Tianjin and Changzhou.

While the early cohort was dominated by urban children of cadre families, there was an attempt to shift enrollments in favor of children from rural and nomadic families. Students were selected on the basis of examinations, according to quotas set for each region of Tibet. Host-city *neidi* schools were paired with specific districts in Tibet for student selection. Teachers of Tibetan language and literature, and some management personnel were also sent to the *neidi* schools. The choices of study program, curriculum, subject teachers, and fees were made by each *neidi* school. Over time, more *neidi* schools were added, selection quotas were modified, and rural- and nomadic-student enrollments increased. The proportion of Tibetan students attending regular senior-secondary schools and going on to college and university rose with the expansion of Chinese higher education.

Student Perspectives

We were able to piece together a preliminary picture of life during and after *neidi* schooling by making field visits to Tibet and several *neidi* schools, and by conducting interviews with 180 *neidi* school graduates (60 each in Lhasa, Shigatse, and Nackchu) who had returned to Tibet. We have assigned each interviewee a case number to protect their identities.

At School

Many interviewees recalled difficulty adapting to the weather and food when they started their *neidi* schooling. Those who went to South China, where the climate was hot and humid, seemed to have more difficulty. In some cases, the students became ill or had skin problems. The interviewees also noted that the air was cleaner and clearer in Tibet, and the water in China tasted differently from that in their hometowns. However, all agreed that they adjusted within a month or two. One interviewee said that the local Chinese dialect used by teachers in Shanxi province was hard to follow at the beginning. Others recalled being sleepy because of the physical effect of moving from a high (Tibet) to low altitude (urban China) area. Girls talked about being homesick and how they could not take good care of themselves without their mothers, who had helped them to wash their hair and make braids. Some, especially the ones who had poor Chinese-language skills before attending inland Tibetan classes, recalled their frustration in trying to understand the teachers in class. The biggest problem involved learning a new language: "During the early period there, the biggest difficulty for me was that I listened but could not understand the Chinese that was spoken and did not know how to speak it. It seemed possible that I spent about a half-year or so to adapt to it."

According to the interviewees, the school teachers took very good care of them, and this helped them to overcome the initial adjustment difficulties. One student recalled that his school grouped students according to their home origins and this helped the newcomers cope with school life. Another said that his school had provided milk for breakfast, as a substitute staple for the butter tea that Tibetan students were used to drinking at home. The answers given to the question of adaptation probably illustrate the situation of most students who were experiencing homesickness and the sudden change in climate and language, but who were helped by the caring, parent-like teachers:

> I cried for two weeks because of homesickness, and even asked to be sent back home. But the teachers were very good, like mothers and fathers they helped me overcome my homesickness. We gradually changed and I came to feel quite happy and life became enjoyable.
>
> I felt very uncomfortable at the beginning. First of all, I was young, and everything had been taken care of by my parents. All of sudden, within a month, I had to take care of myself independently. Second, the climate and environment were very different. I was from a small county in Ali and had taken the long way through Xinjiang to Beijing. I felt lost at the beginning. I could hardly speak Chinese, and the teachers at the inland schools were all Chinese. Students spoke Tibetan with one another if they were from the same dialect area. Otherwise they had to use Chinese to communicate.
>
> Because of the weather, I had skin problems...I was very young and didn't know how to wash clothes and hair etc. I used to be the top student in my primary school. When I was away from my parents' supervision, my school work deteriorated. But gradually, I adapted to the new life well.

All respondents agreed that when they entered the *neidi* schools, the rules were very strict. Moreover, they seemed to appreciate the rules, especially when they reached senior-secondary school where the rules eased up, making them feel lazy. They also commented that their daily schedules were quite regimented. They could only leave the campus on Sundays and had to be accompanied by classmates. Even then, they were required to sign out and report where they spent their time. It was on these Sunday shopping expeditions that they spent the money sent by their parents, money that was stored by their specially assigned teacher.

These teachers were viewed as caring and responsible people. Respondents commented that they would like to treat their own students in the same way when they become teachers. Moreover, they felt they were treated with more respect by *neidi* school teachers than by most teachers in Tibet. Finally, they viewed their *neidi* teachers, especially the *banzuren* (teacher with special responsibility for each class), as a kind of *bangyang* (model to emulate). While they also had a Tibetan teacher at *neidi* school for language and literature class, that teacher usually only stayed for two years and was not as integrated into the teaching network. Nevertheless, respondents agreed that if their Tibetan teacher was skillful, they greatly enjoyed studying Tibetan language and hearing stories from Tibetan literature, especially those that carried moral lessons. Most Chinese *neidi* teachers had never been to Tibet and had little knowledge about it. For example, a student commented that one *neidi* teacher had a stereotypical view of Tibetan students and fear during the first meeting with them.

Respondents also said they learned a great deal from the Han students, despite the fact that the classes, and most of the schools, were ethnically segregated. Most interactions with Han students occurred at sporting competitions, on field trips to the theater, and on special days for joint academic events. Tibetan students viewed Chinese students as very hard working and thought they had better study methods, better learning skills, and a broader vision of the world.

During Tibetan holidays, leaders from Tibet would visit the *neidi* schools. The school would prepare special Tibetan foods. Students also wore their traditional dress and sang Tibetan songs. During long holidays, field trips around the country were arranged for them, from which they felt they learned a great deal. When they did travel, people took note of their behavior, including their speech and appearance. A Tibetan student noted what he perceived as an unwelcome reaction by people in Shanghai. In general, it appears that the wide variety of interactions and experiences that Tibetan students had with their Chinese teachers, canteen workers, off-campus shopkeepers, and others they met on their tours actually strengthened their identity as Tibetans. Although the interviewees sometimes commented that they did not know a great deal about Tibetan history and culture, or that they wanted to learn more of Tibetan literature, there was never any question of confusion about their ethnic identity. For example, there was never a case of a student referring to another as having "converted" from being Tibetan to being Chinese, though they learned in school that one was an ethnic, and the other a national identity. In short, there was a clear expression of Tibetan ethnicity, and it was not discouraged by the school authorities.

The school architecture and environment were similar to others schools in China except that the schools and campuses included Tibetan sculptures, murals, ceramic displays, paintings, and photos. Except for photos of the Panchen Lama, the display of religious symbols in school was not permitted.

The students often wrote letters home, though they have recently begun to use telephones more frequently to call home. At first, the majority wrote letters in Tibetan language because this is the language they knew when they began at *neidi* schools. Later, they would write in Chinese to show their parents what they had learned, even though many parents could not read Chinese and had to find someone to translate the letters. Even when letters written in Tibetan arrived, parents who were illiterate had to find someone who could read them. In one case, a parent who could not read Chinese insisted that his son write in Chinese because it would be a source of pride for the family in the local community. Students generally refrained from telling sad news to their parents, including when they were ill. Parents did likewise. In one case, a girl did not learn of her father's death for three years.

A central issue surrounding *neidi* schools has been the medium of instruction. About 70 percent of the first *neidi*-school cohort was recruited from Tibetan-medium primary schools. The first quotas permitted Lhasa schools, where Chinese was used more often in education, to supply more than a third of the total of all students admitted to the *neidi* schools that year. A 1984 *neidi* school regulation states that Tibetan language is to be the teaching medium in junior-secondary school, but replaced by Chinese language in senior-secondary school. This was unworkable since most of the subject teachers were Chinese. Mathematics, Chinese language and literature (*Hanyu wen*), and Tibetan language and literature (*Zangyu wen*) are three main subjects in the school curriculum. In fact, Chinese language has now become the main teaching medium in all *neidi* schools including at the junior, but especially senior, levels. Students are also required to study a foreign language, which invariably means English.

English language was not originally a compulsory subject but was later made so. Tibetan history and cultural tradition are still not heavily emphasized, despite the 1988 State Education Commission notice that schools in and for Tibet should facilitate the inheritance and development of Tibetan history and cultural tradition, and also learning of advanced scientific technology. The State Education Commission also suggested that the educational content, textbooks, and curriculum design for the Tibetan children should not indiscriminately copy the experience of schools in the place where the *neidi* schools are located. It is suggested they should take into account Tibetan history, culture, production, and economic life. The 1988 notice also prescribed that *neidi* schools strengthen the instruction of Tibetan language in the curriculum. Nevertheless, most *neidi* schools have simply followed the curriculum of the urban schools of the host city where they are located. Consequently, more careful attention to parts of the curriculum concerning Tibetan history, geography, and culture was proposed again in a 1993 Educational Support for Tibet Work Conference. However, our interviews with students indicate the effect was minimal.

Returning to Tibet after Graduation and Readjustment

Virtually all the *neidi* graduates returned to Tibet. Only 5 out of 180 interviewed said that they wanted to stay in inland China to work. Those who stayed to pursue higher education said they also wanted to return to Tibet after graduation. Many interviewees said that they wanted to contribute to the development of Tibet. They were taught the idea of "studying hard now and serving Tibet in the future" by their teachers in the inland Tibetan classes. The monitor teachers, the teachers who took care of the students' daily life, and the Tibetan language teachers had all talked with the students about making a contribution to Tibet after graduation. These conversations took place during the self-study time, and in and after the classes.

The visiting Tibetan leaders always stressed that the purpose of the *neidi* classes was to train Tibetan personnel to work for the construction of the new Tibet. An interviewee recalled that Panchen Lama had visited the school in Kunming and requested that the students study for the rise of Tibet. All such talks made a significant impact on the students. Many spoke of their plans to use the knowledge they learned in inland China for the betterment of Tibet. Some interviewees said that they wanted to go back home to attend to their aging parents, and that they missed their parents and families a great deal and wanted to rejoin them. One put it this way: that "Tibet would always be his spiritual home."

Some respondents said that although they stayed in *neidi* for many years, the environment made them feel isolated and that they never felt at home in *neidi*. As one put it, "At that time, I missed Tibet, and did not want to stay there [in *neidi*]." Another mentioned that even though he stayed in *neidi* for many years, he still disliked the weather, missed home a lot, and wanted to return and make a contribution to Tibet. Others said they knew the people and the environment in Tibet better, so felt it was natural to return to home.

There was also a sense that because the government had paid for their educations, they felt obliged to repay by working in Tibet. They also said that Tibet had a greater need for trained people than China. The students who attended the special programs of *weipei* or *dingxiang,* which were funded by the government and designed to train people in certain skills, were required to return to work in Tibet. Several students among the *weipei* and *dingxiang* students were familiar with the policy that, as they put it: "Where you come from is where you return to." In fact, all interviewees were expected, and/or required, to return to the TAR. They were all assigned jobs in Tibet upon graduation. In two cases, the graduates actually had a chance to stay and work in *neidi*. One mentioned that his teachers asked the graduates to stay and teach the Tibetan students, but no one stayed because everyone was anxious to return home and rejoin their families. Another had joined the Communist Party in the school and the department of Tibet and Xinjiang Affairs Office, under the Ministry of Public Affairs, wanted to recruit him. However, he did not want to stay in *neidi*, because there were very few other Tibetans in *neidi*.

Most interviewees said it did not even occur to them to stay in *neidi*, and they did not find it necessary to give any reasons. For them, it was just very natural to return home. This contrasts with the case of students from the newly established Xinjiang *neidi* schools for Uyghur and other ethnic groups from the Xinjiang Uyghur Autonomous Region. Research indicates that these students would prefer to stay in inland China rather than return to their home provinces.

Most thought that they could not compete in the job market with Han students, so staying in *neidi* was not an option. Language barriers, unfamiliarity with the environment and the people outside of the *neidi* schools, and academic inferiority to the Han students were cited as reasons. There was an acceptance of the idea that, in fact, the *neidi*-school graduates were not prepared adequately to stay in *neidi*. It was impossible for them to think about staying and actually looking for jobs, so there was no choice.

A biology teacher in Shigatse said that he wanted to stay in *neidi* but did not believe he was competitive with the Han students in terms of language skills and subject knowledge. Others might have wanted to stay but felt an obligation to take care of their parents. A few were very specific about reasons for possibly wanting to stay in *neidi,* saying the environment was better and the people were more hard-working. As one put it, his "heart was still in *neidi* China." He read the daily press from *neidi* and kept in touch with people in Shanghai to keep up with developments in his field. It was not uncommon, however, for graduates of the *neidi* schools to consider returning to further their education. Some wanted to study for a tertiary level technical college, but such program did not exist in the early times of the first group of cohorts of neidi students.

Although almost all students wanted to return home to Tibet after graduation, they sometimes faced a long period of adjustment after being away for seven years or more,

> After the *neidi* students returned to Tibet, we eventually came to feel adjusted to life in Tibet. After returning to Tibet we got a strong feeling about returning and belonging although we studied outside. After we were able to see other cities and their development in *neidi* and the things I studied, I felt the *neidi* students had great enthusiasm and ambitions for when we returned [to Tibet]. We felt we did not give it much thought while we were studying for junior secondary and senior secondary. But, after taking the exams for university and graduating from the university, we had great ambitions because we felt we had to do something.... Many of my classmates felt this way. However, there possibly was a mismatch between our specializations and the work we were given. Sometimes, we were assigned to jobs that we did not like, and there was a possibility that we were out of practice in those areas. But, it still was a feeling of returning to Tibet for work, adjusting and having a feeling of belonging.

Recalling what he had missed while away for many years, one replied that he had to relearn many things.

> I will not say it was difficult for me to adjust, because there isn't a big change after all. But, sometimes you realize that you left to study at a young age when you did not know much, and so when you returned to work here, there would definitely be many things that you missed. We are gradually picking up what we have missed. We see, understand and select from these things from our own perspective. That's it.

How long did it take to completely adjust to life back in Tibet? Some said that it took them a month to overcome the uncomfortable feelings from the high altitude and change back to Tibetan food, which contains more meat and fewer vegetables. But others also said that living in Lhasa was not very different than living in any inland city, especially with respect to food and clothes. The ones who went to more remote areas such as Naqu said that they had experienced a mild feeling of high-altitude pressure, and one even had diarrhea because his stomach was used to the Chinese food. Some said it took months, or even a year or more, to completely readjust.

The majority of those interviewed said that, upon returning to Tibet, they did not reject any Tibetan cultural practices. This question was asked in several different ways because the interviewees' had differing understandings of "culture." When asked what aspects of Tibetan society they disliked, some cited poor hygiene habits, and excessive drinking by some people. Others said that older Tibetans, and some Tibetans who had never traveled outside the country, had conservative attitudes toward new things.

However, a large group talked about their resistance, after returning to Tibet, to what they saw as superstitious aspects of Tibetan religion. As one put it, "some Tibetans are superstitious. I don't accept those aspects of religion." Others talked further about their views on Tibetan culture and what should be preserved or discarded:

> Regarding Tibetan culture, it is mainly traditional customs that should be preserved, but the attitude (or view) needs to be changed. For example, feudal superstitious should be changed.
>
> In terms of traditional customs, I feel that some Buddhist traditions have definite influence, like visiting a monastery. In *neidi*, sometimes I did visit some monasteries or temples in Kaifeng (a city in the *neidi*). I went there with our teachers as a tourist, but not to worship Buddha. This is a big change for us. In Tibet, it is now the same for us. When our parents or relatives visit a monastery to worship, we do not go with them. We can take some Han friends to a visit monastery as tourists. I feel this is fine. However, I did engage in the superstitious practice of worshipping Buddha when I was a child. Once I understood this issue historically, I felt there were no such things existing in the world, that they were superstitions. Now I do not believe in these Buddhas, deities or ghosts. I believe some of them existed in history as persons. King Songtsen Ganpo and Princess Wencheng were placed in the monasteries and worshiped as deities, but I felt they were just historical figures and not deities. Therefore, there is a definite change for me in terms of how I see Tibetan Buddhism. Regarding customs, I also have my own thoughts... Some people are begging for food, and some people are throwing *tsampa* (barley flour) everywhere because it is a ritual offering custom. From the scientific and humanistic point of view, this is not right.

There were cases where interviewees said that they had lost their religious beliefs entirely. For example, one former student came from a village near Sagye, one of the oldest and most influential Tibetan monasteries. Before he went to the *neidi* school, he spoke no Chinese. He said that the biggest change in him was that he no longer had religious beliefs. His family members were still believers, but he viewed this as their choice. Otherwise, his identity was still Tibetan and he did not want to change that.

> The biggest change in me is that I do not believe in Buddhism any more. Of course, I am never against other people believing in Buddhism, like my father, mother, older brother, and younger sisters. They all believe in Buddhism, and I am never against them. Buddhism is a psychological medicine for some people. It is a medicine for one's heart. Therefore, I am never against people who believe in it. Of course, I myself would not believe in it. This is a big change in me. When I first went to *neidi*, I believed in religion very much. At that time, my family invited monks for religious rituals predicting my future, and

> did everything like this for me. Now I do not believe in these things. This is the main change. On the other aspects of the Tibetan culture, there is no change in me and I do not want to change. I am still a Tibetan.

Others who joined the Communist Party also said that they no longer believed in religion. This loss of belief does not affect their Tibetan ethnic identity, and they still regard the religion as part of the overall culture. They claimed that their attitude toward religious behaviors changed and became more selective, that is, they still believed in Tibetan religion, but had some resistance toward what they called "superstitious" behaviors of the other Tibetans, but at the same time still maintained Tibetan ethnic identity. "Regarding religion, I have the attitude that there are two sides to everything. Honestly speaking, I am a member of the Communist Party, and it is not possible that I believe in religion... Our Tibetan nationality has 'ten great cultures'... They have taken very strong root in Tibetan Buddhism." Generally, the interviewees are proud of their own cultural heritage. The experience of studying at a *neidi* school, and living in China, reinforced their Tibetan cultural identity and they wanted to work for the homeland.

> I don't think that I resisted Tibetan culture..., especially that of our *neidi* classmates... After staying so long in the mainland, instead of resistance, I have a strong love toward Tibet as my homeland, and I feel I definitely need to do something for her. Concerning Tibetan culture I feel I accept and love most of it, and I am very proud of Tibetan culture.

The students who were proud of Tibetan cultural heritage wanted to learn more about it, but found themselves constrained by their poor skills in the Tibetan language. As one simply put it: "I want to learn about Tibetan culture."

Conclusion

It is clear that the *neidi* schools differ from the stereotype of American institutions designed to erase Native American cultural identity. The *neidi* schools themselves have not been used to deracinate students by prohibiting the use of native languages or facilitating the erasure of memories of traditional culture. Tibetan families are not coerced into sending their children to *neidi* schools. Moreover, some families, whose children do not have scores high enough to gain entrance to these schools, will pay the extra fee for admission or, if they are wealthy enough, will send their children to the growing number of private (*minban*) secondary schools in China.

While the mission of the *neidi* schools may be to integrate Tibetans into Chinese civilization, the schools recognize Tibetan culture by allowing examples of Tibetan art, architecture, and music, and the observance of Tibetan holidays. Nevertheless, many teachers and school principals still hold the view that Tibetan culture is backward and far from the civilized culture of Han China. Tibetan students may be affected by this, but they are also required to study Tibetan language and literature in *neidi* schools. Still, the emphasis on the study of Tibetan language and culture become transitory as students begin to prepare for the national college and university entrance examinations in senior secondary school. The attention of students and *neidi* schools to Tibetan language and literature decreases sharply as the preparation for the national examination intensifies.

The *neidi* schools contrast with boarding schools for indigenous peoples in North America. For example, corporal punishment is not used to force students to speak Chinese while inside or outside of the school, or to control other forms of behavior. Fear is not used to control behavior, though the school often uses moral and political arguments, and teacher-modeling to shape behavior. Communication with parents is not cut off at school, and parents are permitted to visit the schools and increasingly do so. Finally, *neidi* school graduates testify that the relationship between Tibetan students and their Han Chinese teachers is generally positive. In short, there is little evidence that these *neidi* schools resemble the notorious boarding schools for indigenous peoples in Australia, Canada, and the United States that were established to de-culturate.

Many respondents commented that, unlike their counterparts who did not leave Tibet for study in inland China, the *neidi* schools made them more independent and self-reliant. They also believe that studying in the *neidi* schools made them more adaptable to different situations and environments.

Nevertheless, *neidi*-school graduates did find themselves feeling that they did not know enough about their own language, history, and culture. Their Tibetan-language skills had declined and, in many cases, their skills in reading and writing Tibetan were inadequate for life back in Tibet (Postiglione et al. 2007). As institutions for transmission of Tibetan cultural heritage, the *neidi* schools may only play a token role. They could hardly do more since virtually all *neidi* teachers, except their Tibetan-language teacher, have never been to Tibet. After returning to Tibet, students usually require a period of readjustment to Tibetan life. Nevertheless, the *neidi* schools do not diminish Tibetan identity. If anything, they strengthen it by situating classmate networks of young Tibetans within urban Chinese culture for many years.

As for the school curriculum, its mission is to make young Tibetans more Chinese by socializing them into the national mainstream of ways of thinking, feeling, and acting (Bass 1998). Yet, this does not necessarily amount to assimilating them. In rethinking assimilation, for example, Alba and Nee (2003) point out that assimilation into the mainstream is not the only possible form of assimilation. It does not have to mean that the minority changes completely while the majority remains unaffected. While Alba and Nee's definition of assimilation includes a decline in ethnic distinctions, members of the minority groups do not sense a rupture from familiar social and cultural practices as they participate in the mainstream institutions. Furthermore, the mainstream is likely to evolve in the direction of being influenced by members of ethnic and racial groups that were formerly excluded. As *neidi* school graduates become completely accepted as residents and workers in the host cities where they study, and as teachers in other host-city schools, the goal of cultural integration will have been more fully accomplished.

Notes

This chapter acknowledges the support of the Hong Kong Research Grants Council (HKU 7191/02H).

1. There has been a great deal of confusion about the translation of *Xizang neidiban* from Chinese to English. The government's former translation of "Tibet Inland Schools and Classes" could give the impression that these schools and classes are held within Tibet. A more recent government translation is Hinterland Schools. In actuality, the *Xizang neidiban* are largely located in major Chinese cities and are basically Chinese Boarding Schools, thus the more English translation would be dislocated schools.
2. *"Concerning Attaining the Target of the Formation of Interior Region Tibetan Schools and Classes for Cultivating Talented Students,"* Central Government Document Number 22 of 1984; and *"Circular Concerning Attaining the Central Implementation Target of Cultivating Tibetan Talent in the Interior Regions,"* Document Number 25 of 1984.
3. Some are converting to cater to only senior secondary school students from Tibet.

References

Alba, Richard, and Victor Nee. 2003. *Remaking the American Mainstream: Assimilation and Contemporary Immigration*. Cambridge: Harvard University Press.

Bass, Catriona. 1998. *Education in Tibet: Policy and Practice since 1950*. London: Zed Books.

"Concerning Attaining the Target of the Formation of Interior Region Tibetan Schools and Classes for Cultivating Talented Students," Central Government Document Number 22 of 1984.

"Circular Concerning Attaining the Central Implementation Target of Cultivating Tibetan Talent in the Interior Regions," Document Number 25 of 1984.

Fei, Xiaotong. 1986. "Zhonghua minzu de duoyuan yiti geju" [Plurality Within the Organic Unity of the Chinese Nation], *Beijing daxue xuebao* 4: 1–19.

Gladney, Dru. 1995. "Economy and Ethnicity: The Revitalization of a Muslim Minority in Southeastern China," in *The Waning of the Communist State: Economic Origins of Political Decline in Hungary and China*, ed. Andrew Walder, 242–266. Berkeley: University of California Press.

Goldstein, Melvyn C. 1989. *A History of Modern Tibet 1913–1951: The Demise of the Lamaist State*. Los Angeles: University of California Press.

———. 1997. *The Snow Lion and the Dragon: China, Tibet, and the Dali Lama*. Los Angeles: The University of California Press.

Irendale, R., Naran Bilik, Wang Su, Fei Guo, and Caroline Hoy. 2001. *Contemporary Minority Migration, Education and Ethnicity*. Cheltenham, UK, and Northampton, MA: Edward Elgar.

Ma, Rong. 2007. "Bilingual Education for China's Ethnic Minorities," *Chinese Education and Society* 40 (2): 9–24.

Mackerras, Colin. 1999. "Religion and the Education of China's Minorities," in *China's National Minority Education: Culture, Schooling and Development*, ed. Gerard Postiglione, 23–54. New York: Falmer Press.

N.A. 2007. "Fourteen Thousand Talents were Nurtured by the Inland Tibetan Classes in the Past 20 Years," *Xizang Ribao* [Tibet Daily], January 29.

Nyima. 2000. *Wenming de kunhuo: Zangzu de jiaoyu zhilu* [The Puzzle of Civilization: The Way out for Tibetan Education]. Chengdu: Sichuan minzu chubanshe.

Postiglione, Gerard, Ben Jiao, and Manlaji. 2007. "Language in Tibetan Education: the Case of the *neidiban*," in *Bilingual Education in China: Practices, Policies and Concepts*, ed. Anwei Feng, 49–71. New York: Multilingual Matters.

Postiglione, Gerard, Zhiyong Zhu, and Ben Jiao. 2004. "From Ethnic Segregation to Impact Integration: State Schooling and Identity Construction for Rural Tibetans," *Asian Ethnicity* 5 (2): 195–217.

Sautman, Barry. 1999. "Expanding Access to Higher Education for China's National Minorities: Policies of Preferential Admissions," in *China's National Minority Education: Culture, Schooling and Development*, ed. Gerard Postiglione, 173–211. New York: Falmer Press.

Xizang. 2005. Xizang zizhiqu zhi: Jiaoyu zhi [Chronicle of the Tibetan Autonomous Region: Education]. Beijing: Zhongguo zangxue zhongxin.

Zhang, Tianlu. 1997. *Population Development in Tibet and Related Issues*. Beijing: Foreign Language Press.

Zhou, Minglang, and Hongkai Sun (eds.). 2004. *Language Policy in the People's Republic of China: Theory and Practice Since 1949*. Norwell, MA, Kluwer Academic Press.

Zhu, Zhiyong. 2007. State Schooling and Ethnic Identity: The Politics of a Tibetan Neidi School in China. New York: Lexington Press.

Chapter 8

School Consolidation in Rural Sichuan: Quality versus Equality

Christina Y. Chan and Stevan Harrell

In the year 2000, Yanyuan County, a poor, mountainous minority area in southwestern Sichuan, consolidated its elementary school system. The county closed 90 percent of village primary schools and expanded "key-point" schools located in townships and county seats. Consolidation was meant to reallocate education resources in response to greater financial pressure on the local government and a rising number of failing village primary schools.

This study looks at the recent consolidation and its impact on the provision of education in rural areas.[1] We examine the situation of local education both through aggregate data and through comparison of five elementary schools. Each field site is differentiated not only by administrative level (county, township, or village) but also by its position in the county's economic structure, using a model loosely based on G. William Skinner's model of hierarchical regional space (HRS; Skinner, Henderson, and Yuan 2000).

The newly consolidated Yanyuan County elementary school system has had many triumphs in recent years: enrollment has reached record highs (from 25,057 primary students in 1990 to 40,352 in 2003), and there is now one teacher for every twenty students. However, the aggregate time-series data do not reveal how educational benefits are distributed within the county. Government data use the term "rural" to blanket agricultural regions at the sub-county level, thereby hiding emerging disparities among townships and villages. We find that within Yanyuan County, although the school consolidation policy has increased the average quality of basic education,[2] schools have become less accessible to students living in remote areas. In addition, the few village schools remaining after the 2000 consolidation have had to find other sources of funding in order to maintain school facilities and quality.

Villages situated in the vicinity of key-point schools reap the benefits of consolidation. Yet other villages are being left out. With poor infrastructure connecting villages with township centers, the new consolidation policy exacerbates the polarization between remote villages in the county periphery and developing areas in the county core.

China's Rural Education Post Mao

Chairman Mao believed in having "one school in every village." During the Cultural Revolution (1966–76), the education system was completely decentralized (Mauger 1983; Pepper 1981). The curriculum was geared toward political and moral indoctrination, a complete reversal of the science-based system established by the Republican and Communist Governments from the 1930s to the 1960s. After Mao's death in 1976, leaders of the Chinese Communist Party (CCP) hoped to close the development gap between China and other first world countries by fostering a new generation of scientists, engineers, and academics. The education bureaucracy was re-centralized in Beijing, and competitive entrance examinations were once again established (Pepper 1981).

The central government has identified impoverished rural areas as the weak link in China's education system. The success of top-down reforms, such as the *Nine-Year Compulsory Education Law* (1986), is strongly correlated to the level of development of an area (Hannum 1999). In rural areas high transportation costs greatly impede access to education. Rural inhabitants are scattered across the countryside, resulting in a disconnected system of village primary schools with few resources (Pepper 1990).

Fiscal decentralization has also aggravated regional inequalities by forcing revenue-starved rural counties to become self-sufficient. Such areas may choose to divert investment from public infrastructure to revenue-gathering industries such as Township and Village Enterprises (TVE; Park et al. 1996). This means possible delays in wages for teachers and decreases in school maintenance funds.[3]

Without being able to rely on adequate and timely support from the local government, some schools have had to find creative ways to raise funds. In the Shiyan Municipality, an impoverished area of Hubei, villagers are encouraged to donate profits from crops to the local school (Tsang 1994). Schools may also employ private fees to cover *minban* teachers' salaries, books, and maintenance. *Minban* teachers, most recruited in the 1970s and not formally trained, are not on the government payroll. The need for rural schools to be somewhat self-sufficient contrasts with the state-funded status of urban schools, raising the question: "Can China ever realistically offer equal education opportunity to inhabitants of remote towns and villages?"

Yanyuan County's Education Situation

Yanyuan County is located in the Liangshan Yi Autonomous Prefecture in a region known as Xiao Liangshan (Lesser Cool Mountains) on the border of southwest Sichuan and northwest Yunnan provinces (see figure 8.1). The county is centered on a broad basin of about 2,400 meters elevation, surrounded by mountains as high as 4,300 meters and deep river valleys as low as 1,060 meters. The basin flatland, the most densely populated area, has decent transportation infrastructure with paved roads and buses that run every ten–fifteen minutes. In contrast, the mountain townships and villages are at best accessible by muddy dirt roads and in some areas only via foot or motorcycle.

There are several ethnic groups that inhabit the area. Yanyuan's population is made up of 45 percent Nuosu,[4] 47 percent Han, and 8 percent Prmi and Na. The Han Chinese live mostly in the basin and in some of the river valleys, while the Nuosu villages are located primarily in the mountains. Like many mountainous areas, especially in Western China, the region is quite poor due in part to its geographic remoteness. Within the county, Nuosu areas are generally poorer and have worse infrastructure, including schools, than do the Han areas in the central basin.

Before consolidation, most villages had a small schoolhouse, but conditions were poor.[5] The typical schoolhouse was a one-room building made out of mud bricks. Moreover, education resources were not distributed based on demand; some classrooms had sixty students under one teacher while others had fifteen students under three teachers.

Teachers are the most limited education resource. Most teachers, even those who themselves grew up in small villages, want to teach in either a developed township or the county seat. This is a reflection of China's social climate: after having received an education and been certified by the government, trained teachers do not want to return to their previous standard of living.[6] Even well-funded schools in small villages are unattractive because life in these villages revolves entirely around the school; there are no other opportunities for recreation. County data indicate that there exist an adequate number of teachers in Yanyuan. Figure 8.2 shows Yanyuan's county-wide student-to-teacher ratio to be above the government's preferred standard (one : twenty).[7] Yet some rural schools still report a shortage. The problem therefore lies in the distribution of teachers within the county.

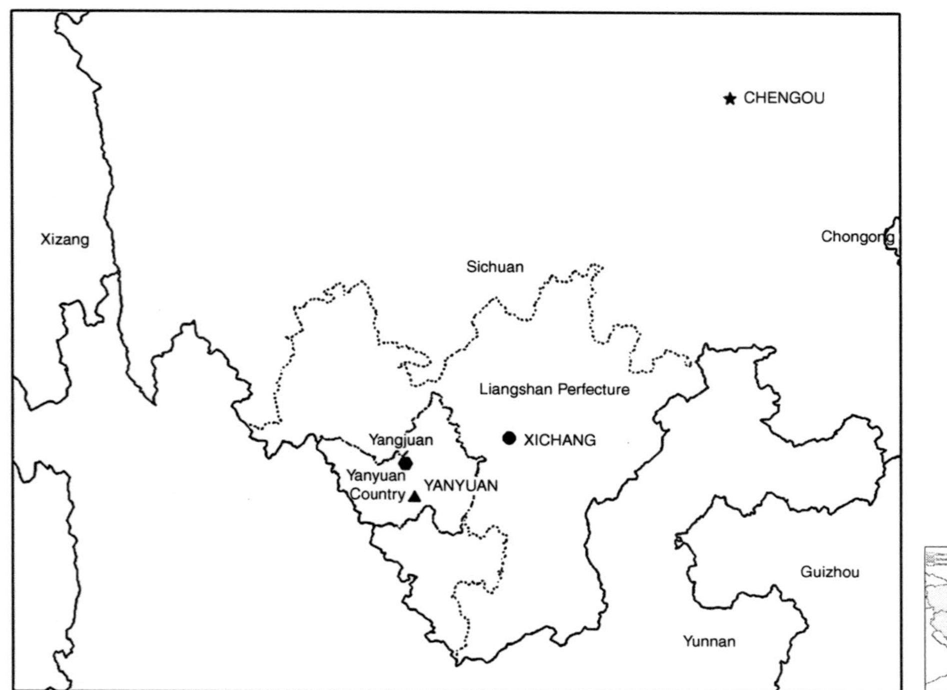

Figure 8.1 Map of Liangshan Prefecture and location of Yanyuan County in Sichuan Province. Inset: Map of Sichuan Province.

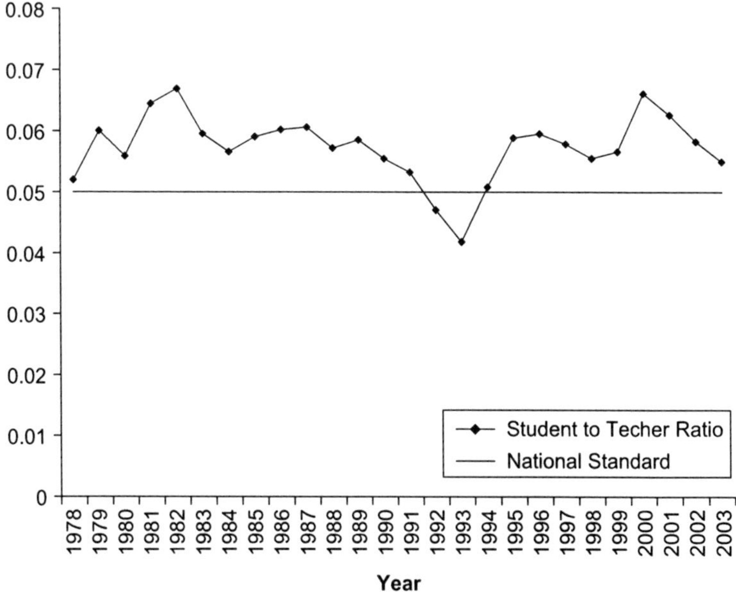

Figure 8.2 Yearly student to teacher ratio in Yanyuan County.

The poor quality of village primary schools, including the shortage of teachers, has impacted the demand side of education. Without an adequate number of teachers or facilities, students are unable to compete in entrance examinations for secondary schools, resulting in low returns on education. Families who wish their children to attend school prefer sending them to higher quality schools in developed townships. Therefore not only do village schools suffer from underfunding, they also often suffer from under-enrollment.

By the end of the century, it had become apparent to education officials from both the provincial and the county levels that the system of scattered primary schools was not adequately serving the rural population. In the year 2000, Yanyuan County, along with many other rural counties in Sichuan, followed national policy and conducted a massive consolidation of primary schools. The primary education system was to be modeled after the secondary education system, that is, a system of well-funded key-point schools that serve as magnet schools for the surrounding area. Village-level schools were to be closed and key-point schools built in townships and the county seat. By 2001, 66 percent of all elementary schools within the county had been closed (see figures 8.3 and 8.4).[8]

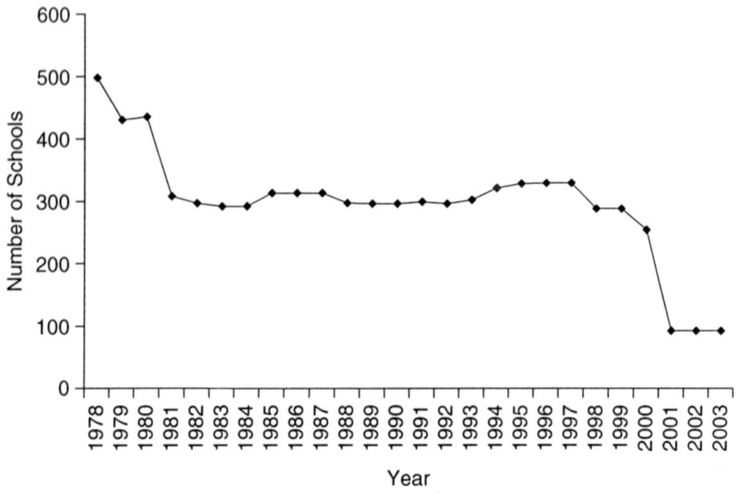

Figure 8.3 Official yearly count of primary schools in Yanyuan County.

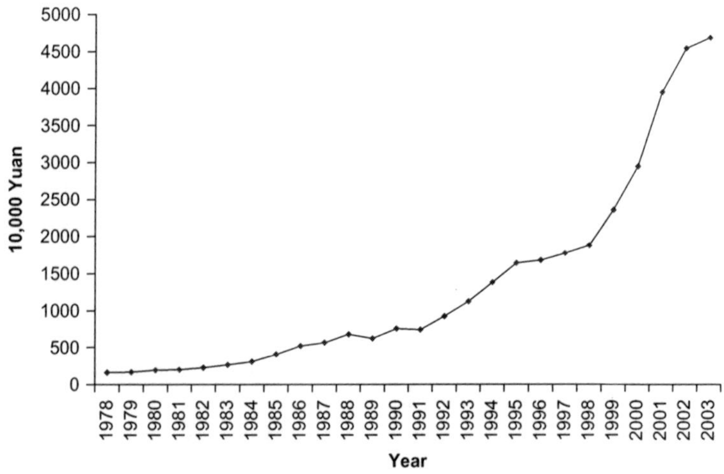

Figure 8.4 Adjusted yearly expenditure on education in Yanyuan County.

Although enrollment is at an all-time high, the county must now ensure that the opportunities for education are equally accessible. The gross numbers depicted in figure 8.5 do not reveal where the increases in the number of students come from. The increases are likely to come mostly from economic centers or the central basin. Resources and opportunities are

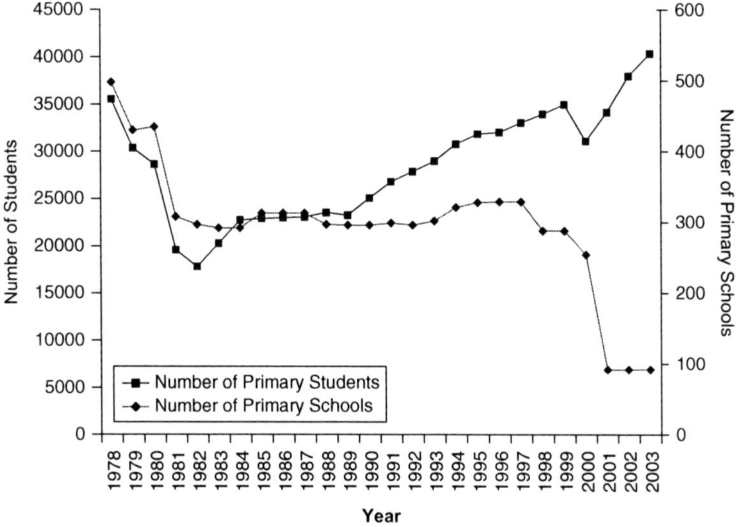

Figure 8.5 Number of schools versus enrollment in Yanyuan County.

clustered in certain areas, increasing transportation costs for those who live on the far peripheries of these development centers.

The Effects of Consolidation: A Comparison of Key-Point Schools and Successful Village Schools

Fieldwork was conducted in two key-point schools, Yanyuan Government Street Elementary and Baiwu Township Elementary, and three non-key-point elementary schools in the rural villages of Shaba, Yangjuan, and Mianba.[9] Both key-point schools are in local economic centers but on different administrative levels: Government Street Elementary School (GSES) is an example of a county-level key-point school located in the central basin. Baiwu Elementary is an example of a township key-point school located in Yanyuan's mountainous areas. Both schools have their respective administrative regions from which they draw their students and resources. GSES attracts students from relatively wealthy neighboring townships, while Baiwu Elementary serves as a magnet school for nearby and remote villages under its jurisdiction.

The centralization of elementary education in key-point schools located in economic centers provides numerous benefits. The developed local economies surrounding key-point schools provide diverse sources of funding. These areas may draw funds not only from agricultural production but also from tourism and the small number of primary industries. The top-down nature of the Chinese government means that the county government's information network extends reliably to townships but not necessarily to villages. Village schools must deal with an extra layer of administrative bureaucracy when communicating with the county government.

Yanyuan still has a small number of functioning village schools. Two of the three village schools studied here were able to find independent sources of funds and avoid closing during the 2000 consolidation. Shaba Elementary is located a half-hour drive away from Yanyuan GSES. Shaba Elementary has been receiving corporate funds from China Telecommunications as part of the company's development financing project. Yangjuan Elementary is located in a village within the Baiwu Administrative Township and is about a forty-minute walk from Baiwu Elementary. Yangjuan has benefited from funds raised by foreign researchers and philanthropists. Mianba, another village in the Baiwu Township, has received no domestic or foreign assistance for its school, and is thus perennially on the point of closing altogether.

Students are more attracted to key-point schools than to village schools. Their mindset is that attending a key-point school is equivalent to working toward higher socioeconomic status.[10] Students will often walk two–three hours to attend key-point schools. Also village primary schools are more likely to charge attendance, book, and/or boarding fees due to their need for funds. The county is in the process of implementing the policy of *liang mian yi bu*, or "two waivers and one stipend," that is, eliminating book charges and miscellaneous fees, and helping with living expenses, but many village schools have not totally abolished fees. Therefore, although village schools provide an alternative to key-point schools, they are disadvantaged in attracting students and trained teachers.

Basin versus Mountains

There is also a geographic dimension to the effects of the consolidation. Key-point schools have been built mostly in central basin townships. These townships have the economic resources to support large schools and also have higher population densities. Moreover, the transportation costs for financial and human capital are much lower in the basin than in the mountains. It only takes a bus ride to experience the difference in transportation costs: basin townships have buses that run every ten minutes and travel on relatively smooth paved roads. Mountain township buses run sometimes only once a day and visit only the most accessible villages.

Central basin areas are more likely to attract trained teachers due to the higher standards of living and low transportation costs. Teachers in mountain areas can often only afford to travel to the county seat once a week, while central basin teachers can go perhaps every other day.

The central basin versus the surrounding mountains is thus another dimension we can use to compare schools. Regardless of whether a school is a key-point or a village school, its geographic location has a large impact on what resources are available to that school. Shaba village and the Yanjing County seat are both located in the central Yanyuan basin. Baiwu Township, including Yangjuan and Mianaba villages, is located in a mountainous region north of the county seat, a ninety-minute bus ride away.

Methodology

To determine the differences in access to education between the catchment areas of key-point and village schools, and of central basin and mountain schools, both time-series quantitative data on numbers of schools and enrollments and budgets, and local interview data were gathered. Time-series quantitative data came from a variety of sources, as most compilations were incomplete. The main sources of quantitative data were the Yanyuan County 1990 Book of Statistics, for figures before 1990 (*Yanyuan xian zhi* 1991), and unpublished data kept in Yanyuan County government offices, primarily from the Office of Education and Cultural Affairs. Data for villages came from one-on-one interviews with village heads, school principals, and officials at the township party headquarters.

We conducted several interviews with the county education and tax officials. Local leaders were asked to reflect on the reasons for recent education reforms and their effectiveness. At each field site we met with school principals and teachers. They were asked how the reforms have affected their schools and whether they thought these changes were beneficial for their local community and for the county as a whole.

Delineating the Differences in Development between Selected Schools

The schools in this study were selected because of their location in areas of different stages of development. The hierarchy of development was decided based loosely on G. W. Skinner's HRS model. Skinner's model allows us to get beyond the misleading dichotomy between "urban" and "rural"

Table 8.1 Hierarchical regional space (HRS) model for comparing field sites*

CPZ		County seat	Developed township	Market town	Agricultural output center
URC		1	2	3	4
County	1	Yanjing (2)			
Township	2			Baiwu (5)	
Village	3		Shaba (5)		Yangjuan, Mianba (7)

*Number in parentheses denotes HRS index. Higher numbers denote lower levels of development. CPZ: Core-periphery zone; URC: Urban-rural continuum.

(Skinner et al. 2000). In recent research, Skinner's model is applied mostly on a macro-geographic level, identifying major cities such as Shanghai as apex metropolises and then classifying other towns and cities in relation to these apexes in order to compare levels of development. Skinner determined an area's HRS index by a seven x eight matrix. One variable was an eight-level urban-rural continuum (URC) and the other variable was a seven-level macro-regional zoning of an area's geographic location relative to a core economic zone (called CPZ, meaning core-periphery zone).

Skinner's model is used here on a micro-geographic level to examine the economic development and resource distribution of a particular county. Under the assumption that administrative level is highly correlated with development, we simplified the URC administrative levels into three categories: county, township, village. The CPZ is similarly simplified to county economic center, developed township, market township, and agricultural output area (see table 8.1).[11]

A Description of Field Sites

Yanyuan's Government Street Elementary—County Seat School in High Demand and Key-Point School in the Central Basin

Yanyuan's GSES is located directly in the center of Yanyuan's administrative district in Yanjing, surrounded by government officials' houses and administrative buildings. It is one of three elementary schools located in the county seat. It serves a population of around thirty thousand. GSES is known for having the best teachers in the county, and it also has the lowest number of *minban* teachers of any of the schools sampled in this study.

As of early 2006, GSES served eleven hundred students, half of whom traveled from outside the county seat town of Yanjing. The ethnic composition of the student body is 60 percent Han Chinese, 35 percent Yi, and 5 percent other minority groups.[12]

GSES is overenrolled and its location in an urban area allows no room for expansion. Currently there are sixty–seventy students in one classroom. Other schools in this study, such as the elementary schools in Yangjuan, Baiwu, and Shaba, are also facing the same problems with over-enrollment, but unlike Yanjing's school, their problem lies in the lack of teachers and not in the lack of space.

Shaba Village–Benefiting from Corporate Funding; Meiyu Township–Village School in the Central Basin

The village of Shaba lies in the developed basin township of Meiyu, a half-hour bus ride away from Yanjing town. Shaba makes up the largest administrative village in Yanyuan County, with a population of over five thousand people. The average per capita cash income in 2005 was 1,000 RMB, very high for Yanyuan County. The original school was built in 1956 as a *minban* school, entirely supported by the community. In the 1990s, the county education bureau made plans to shut down the school and transfer the students to the primary school in Meiyu Township.

In the year 2000, a Yanyuan vice-county executive, a graduate of Shaba Elementary, negotiated with China Telecommunications to incorporate the Shaba primary school into its rural development initiative. Although the company representatives originally wanted to give the money to a more mountainous region, China Telecom granted 700,000 RMB directly to Shaba Elementary. The school now runs on the China Telecom funds and its own revenue. The school has a few fields of apples and corn planted by students and teachers; the output is sold and profits are used to pay *minban* teachers and maintenance fees.

Shaba Elementary is now the largest village-level elementary school in the entire county. In 2005–2006 there were 927 students, about a 300 percent increase from the time before the school was rebuilt. The school building is actually capable of holding twenty-two hundred students, but education services are limited because of the lack of teachers.

Baiwu Elementary—Township Key-Point School in the Mountains

The township of Baiwu is the economic and political center for twelve administrative villages in this mountainous Yi region, encompassing numerous local settlements, including the two field sites of Yangjuan and Mianba. Baiwu is a very poor township; its education infrastructure is entirely reliant on county funds.

Despite local poverty, Baiwu's key-point elementary school is relatively well-equipped. The school, established in 1957, serves as the education center for the entire township administrative region. Only two other villages in the township have elementary schools that have all grade levels. Local students must travel up to two–three hours to attend Baiwu Elementary. It currently has fifteen hundred students, of whom 35 percent live at the school.

The size and the facilities of Baiwu Elementary are impressive. The school covers the largest amount of area of any in this study. Since 2000 the school has been expanding; there are now fifteen new classrooms and one new administrative building.

The main challenge Baiwu Elementary now faces is the increasing number of students who wish to attend. The school is experiencing a shortage of teachers; the current student-teacher ratio is far above the ideal of one : twenty students. Also, many students who wish to attend the township school cannot endure the daily commute to school.

Yangjuan Village—The Fortune of Foreign Funding; Baiwu Township—Village School in the Mountains

The villages of Yangjuan and Pianshui are hidden in a valley between the mountains surrounding the small river plain near Baiwu Township. Inhabited entirely by Nuosu people, Yangjuan village is distinct because it receives outside funding from international donors. This was the result of the close contact between Ma Lunzy (Ma Erzi), a native of the village, and a prominent international scholar, and several Chinese and foreign anthropologists and other researchers. At Ma's instigation, some of those foreign associates raised money to build the school, which opened in fall 2000, just as the consolidation policy was eliminating other elementary schools in places such as Yangjuan. Before the local school was built, students from Yangjuan and Pianshui commuted to the Baiwu Township school. Before the year 2000, approximately 26 percent of primary age students attended school, but after the opening this increased to 83 percent (92 percent of the boys and 76 percent of the girls).

The Yangjuan principal believes that his students are worldlier because of foreign influence. The school supports a curriculum beyond basic exam material. In 2006 the top seven grade 6 graduates of Yangjuan had higher test scores than the highest-testing graduate of Baiwu. Due to the unique aspects of the school, Yangjuan Elementary has become a nontraditional magnet school for the surrounding mountainous area, mitigating the problem of over-enrollment at Baiwu Elementary.

Mianba Village—Neither Key-Point nor Outside-Aided; Baiwu Township—Village School in the Mountains

An hour-and-a-half hike from Yangjuan village, through fields of buckwheat and potatoes, is the Nuosu village of Mianba. There is no main road to Mianba, making it almost inaccessible to motorized vehicles. Before 1994, Mianba Elementary provided classes up to the Grade 5. Mianba had a very influential and gifted teacher who attracted students to the school. After 1994 this teacher left the village and the school quality began to decline, exacerbated by the cut-off of government funds during school consolidation. In 2006 the school had only two *minban* teachers. There were thirty-eight students in grade 1 and eight in grade 2—no other grades were taught. The lack of students is by no means due to a lack of demand for education. There are enough students in Mianba to fill the school to its original capacity. Most parents, however, choose to send their children to Yangjuan or Baiwu Elementary.

Key-point and Village School Differences

It is clear that already overcrowded key-point schools would be under even greater enrollment pressure were it not for the existence of local village schools. Village schools help siphon off enrollment by providing an alternative option to key-point schools. Often, a village school will serve not only its own village but also students from surrounding villages who do not wish to commute to the township.

The average number of students who attend key-point schools is still almost double that of village schools. Yanjing GSES and Baiwu Elementary are purposefully equipped by the government to serve over one thousand elementary students each. These schools have larger buildings and also additional facilities such as libraries and laboratories (see table 8.2).

One of the more striking differences between key-point and village schools, other than enrollment numbers, is the percentage of *minban* teachers who make up the teaching staff. Village schools rely much more on community-sponsored (*minban*) teachers. The living standards and location of village schools do not attract trained teachers. The county government will allocate the most inexperienced teachers to teach in villages, but these teachers will attempt to transfer to key-point schools. In interviews with village school teaching staff, many complain of boredom, low wages, and loneliness. One teacher stated that he was getting old but still could not find a wife because no woman would want to move out to the village. *Minban* teachers are unhappy at the village schools where they, on average, receive lower wages than their key-point counterparts. *Minban*

Table 8.2 Comparison between key-point and village schools, Yanyuan County

	Key-point school	Village school
Average number of students	1,300	610
Average ratio of *minban* to official teachers	0.23	0.39
Average salary of official teachers	1,150 RMB	1,000 RMB
Average salary of *minban* teachers	300 RMB	225 RMB
Receiving government salaries	All teachers and staff	Official teachers only
Government funded facilities	Libraries, laboratories, dormitories	None
Other sources of funding	None	China Telecom grants, animal husbandry, foreign funding, farm plots

teachers often have to take second jobs such as chauffeuring, as well as working on their families' farms, because they cannot support their families on teaching wages that the community provides.

Structurally, village schools differ most from key-point schools in their sources of funding. The county government does not make much room in the education budget for funding village primary schools. Key-point schools are entirely supported by the county-level government and also can receive subsidies from the prefectural level. In contrast, village schools rely heavily upon outside funding. Without outside funding, these schools would have had to close under the consolidation effort, or at best limp along like Mianba Elementary. Village schools also have many small revenue-generating projects, such as growing produce or raising livestock. These projects, however, only generate enough revenue to pay for extra textbooks or school supplies and cannot in themselves support a school.[13]

Comparing Basin and Mountain Schools

The comparison data between central basin and mountain schools look similar to those of key-point and village schools (see table 8.3). Basin schools tend to serve more students. The cause-and-effect rationality is unclear

Table 8.3 Comparison between central basin and mountain schools, Yanyuan County

	Basin	Mountain
Average number of students	1,010	900
Average ratio of *minban* to official teachers	0.16	0.47
Average salary of official teachers	1,150 RMB	1,000 RMB
Average salary of *minban* teachers	300 RMB	225 RMB
Total number of boarding students	260	575
Total number of outside students	1,010	200

here; it seems to be a combination of higher population density and higher quality that results in larger schools. The central basin is wealthier and as a result can support larger schools. In Meiyu Township (where Shaba is located) the living standards provide a stark contrast to a mountainous village such as Yangjuan. It is not uncommon for a family in Meiyu to have a concrete-walled house with electricity and telephone. In contrast, the average house in Yangjuan is still in the traditional Nuosu style made out of mud bricks, built around a fire-pit with one single electric bulb hanging from the ceiling, and there are no landline telephones in the entire village.

The central basin serves as the economic center for the county. Farming families travel to the basin to sell their produce. Farmers' children will accompany their parents in the morning to attend the local school. In the schools studied, 50 percent of the student population of central basin schools came from outside the locality, while mountain schools had less than 20 percent.

Mountain schools serve fewer students from outside their administrative areas, despite having a higher number of boarders. Boarders in these schools come from villages that are far away from the township. The only way to reach these villages would be walking for hours or hitching a ride on a horse cart. As a result, families who wish their children to attend the township school will either try to have them move in with relatives who live in the township or have their children board. Boarding used to be an expensive option, but with the new *liang mian yi bu* subsidy, the number of student boarders is limited not by cost but by how much space is available in the dormitories.

Mountain schools are heavily dependent upon *minban* teachers. This is true even for mountain key-point schools; for example, 40 percent *minban* teachers staff Baiwu's key-point school.[14] The underlying causes are the same as for village schools' staffing problems: mountain areas do not offer the standard of living the central basin can provide. *Minban* teachers at central basin schools tend to receive higher wages than their mountain counterparts. In 2006, the *minban* teachers in Shaba made around 300 RMB per month, while those in Mianba made only 150 RMB.

The Skewed Distribution of Education Opportunities

The benefits that households reap due to the changes made to the education infrastructure depend on geographic and economic factors of the region. The Chinese government has concentrated developmental efforts entirely on local economic centers, neglecting the most remote villages.

These inequalities are evident from the case studies presented earlier, but here we reinforce our findings with quantitative data from Baiwu Township, which can serve as an example of emerging geographic disparities in education opportunity (see figure 8.6). As described in the previous section, Baiwu Elementary is a well-funded key-point school, one of the largest in the county. The surrounding villages, except for Dalin, which was formerly a township center before being amalgamated with Baiwu, have had education funds cut as part of the consolidation program. Village children who wish to continue in school either walk to school, stay with relatives, or board. Many, however, simply do not attend school.

Costs of education increase the farther away a household is from the township. As a local teacher described,

> Getting to school for many students means they must wake up in the morning, and before the sun has even risen, take a two to three hour walk to the township. The road goes up and down, and it is not easy. Then in the afternoon they must make the same trip back to their homes. By the time they reach home, in the winter especially, it may already be too dark to study. Such conditions naturally affect the incentives for students to attend school.

Table 8.4 clearly shows the effects of high transportation costs in Baiwu Township. In villages located more than five kilometers away from a key-point or a multi-year village school, elementary school students on average make up about 5 percent of the total number of elementary school students in the township. Villages located near either a key-point or a multi-grade village primary school have on average 20 percent of the elementary students within the population.[15]

A map of the distribution of elementary schools in Yanyuan County shows that schools are clustered around flatland areas or in close proximity with townships (see figure 8.6). This leaves vast areas without convenient access to primary education (see figure 8.7). We define "convenient access" as being located within a five-kilometer radius, a walkable distance in an hour on dirt roads or trails. While some areas have several schools competing for local students and trained teachers, townships on the peripheries remain underprovided.

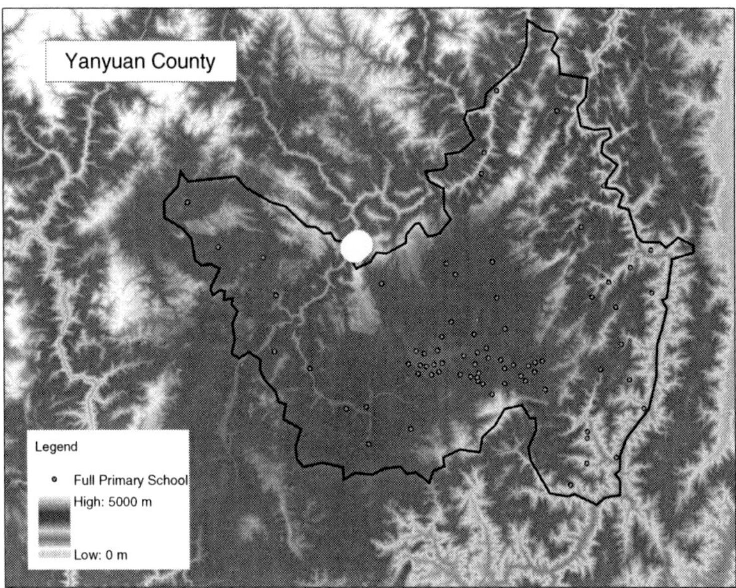

Figure 8.6 Map of location of primary schools with reference to elevation.

Table 8.4 Ratio of primary students to population by administrative village in Baiwu Township, Yanyuan County

Village	Number of primary students	Village population	Ratio of students to population
Dalin*	389	2,853	0.17
Baoqing*	149	868	0.17
Yangjuan*	300	900	0.33
Changma	29	1,401	0.02
Changping	143	2,959	0.05
Mianba	37	1,716	0.02
Zumo	36	1,427	0.03
Maidi	111	1,861	0.06
Shanmen**	90	854	0.11

*Villages have a full primary school (grades 1–6) within a five-kilometer radius. **Shanmen is less than five kilometers away from a primary school with grades 1–4.

Yangjuan and Baiwu Elementary are examples of two schools competing for the same population of students. The two schools are about a forty-five-minutes' walk apart and draw their students from the same areas. Yangjuan was established because of the close personal connections between one of its

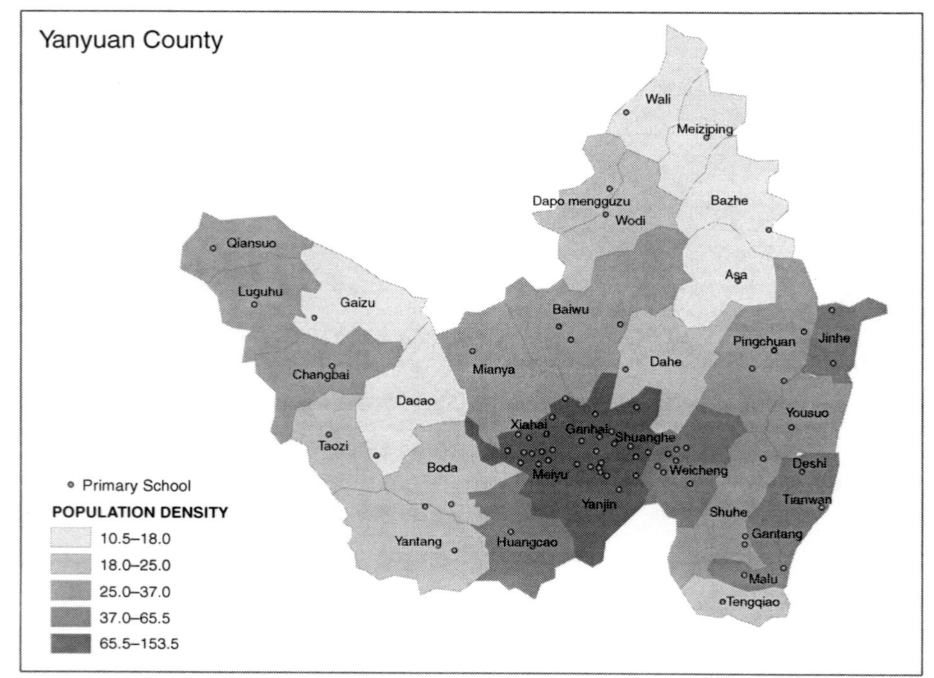

Figure 8.7 Location of county primary schools with reference to township population densities.

natives and foreign scholars; it would have been more efficient to fund a school in a village like Mianba so that the children in the village could avoid the two-hour commute to the township. A more dispersed system of village schools could still mitigate overcrowding in key-point schools while improving access to education among the rural population.

Discussion and Conclusions

Because of the lack of reliable time-series data we cannot measure how education opportunities were distributed in Yanyuan before consolidation,[16] and therefore it cannot be said conclusively that the consolidation effort has resulted in increasing inequality. We can, however, show that rather extreme inequality of access exists at present, and that the potential for greater regional disparities has increased. Limited access to education and higher economic costs to families create disincentives to educate children, although it is entirely possible that potential returns to education outweigh increasing costs. However, in an economy increasingly valuing human capital (Hannum et al. 2008), access to education becomes ever more important for those wanting to escape the cycle of poverty.

The efforts of the Chinese government, particularly the provincial levels, to increase the quality of basic education should be applauded. Some reforms have succeeded, but at the cost of unequal access for children living in remote areas. Higher direct costs to students are mitigated in a few cases by the existence of successful, subsidized village elementary schools. These village schools only exist because of outside assistance, receiving little or no support from the government. The current education system in Yanyuan County relies to some extent on schools that are not actually part of the government planned system; where these schools do not exist, as in most of the villages in Baiwu Township, children are much less likely to receive education. The government's education plan in itself cannot meet the goal of universal mass education because it creates enrollment pressures and high student transportation costs that have been alleviated in only a few cases by outside funding.

Is the Chinese government willing to rely on outside support to serve its education needs? If the Chinese government wants to pursue a purely public education system, there needs to be serious rethinking of the consolidation policy. There need to be more schools in mountainous areas to decrease transportation costs. Moreover, the locations for these schools should not be chosen by economic prosperity or by outside connections of local people, but by whether that location will improve access to education for marginalized groups. Groups of villages in remote areas of the mountains should have their own government-sponsored school as an alternative to the township school. In addition, increased investment in transportation infrastructure in the remote areas could be the key to decreasing costs to students and providing more incentives for teachers to work in mountain regions. Paved roads and regular bus service can cut transportation time by two-thirds or more.

If private philanthropists continue to be an essential part of the education system, there needs to be further effort to provide funds to areas that are in most need. In this study, the schools of Yangjuan and Shaba were built not in the most desperate areas, but where the respective philanthropists had local connections. This is not to say they were unfairly favoring one area; rather, logistical complexities such as official approval and permits made it necessary to have existing connections with local leaders in order to make the bureaucratic process more expedient. There needs to be better communication and coordination between philanthropists, locals, and county governments to pinpoint areas in most need. Perhaps academic institutions could serve as an impartial party to survey rural counties to identify potential locations for new schools. In the meantime, the dream of universal primary education for all China's children continues to go unfulfilled, especially in the remote areas populated by members of minority nationalities.

Notes

1. The authors would like to thank Tami Blumenfield and Ann Maxwell Hill for comments on earlier drafts of this chapter, Amanda Henck for much help with maps, Li Xingxing for introducing us to Shaba, and Ma Vihly for providing all kinds of invaluable help.
2. Quality of schools was evaluated in terms of number of teachers, building maintenance, school facilities, and learning conditions.
3. The dilemma of county officials with insufficient educational funds has been tragically illustrated by the collapse of schools in many areas affected by the Sichuan earthquake of May 12, 2008. Officials caught between the necessity of building a school and the inadequacy of available funds often ended up compromising on shoddy and ultimately deadly construction.
4. Nuosu, numbering about two million, are a subgroup of the official Yi *minzu* (national minority). Most Nuosu live in Liangshan Prefecture and adjacent areas in Sichuan and Yunnan. For overviews of the Nuosu and their educational system, see Bradley 2001; Harrell and Bamo 1998; Harrell and Ma 1999; Harrell et al. 2000; and Schoenhals 2001.
5. Interview with county official from the Office of Education and Culture.
6. Interview with primary school teachers from Yangjuan, Baiwu, and Mianba elementary.
7. Interview with county official from the Office of Education and Culture.
8. In truth, the consolidation of the primary school system had already been slowly occurring since the early 1990s. Throughout the 1990s, many village primary

schools were shut down or downsized due to a lack of funds, teachers, and students. Yanyuan County had been experiencing a decline in the number of schools despite an increase in education funds. Education administrators over the years decided that funds were better used in schools that were already successful while allowing failing schools to close. However, it was not until 2000 that it became official policy to concentrate all funds on key-point schools while providing only basic maintenance funds to other schools.

9. Christina Chan conducted all the fieldwork in these schools except for Yangjuan. Stevan Harrell was a founder of the Yangjuan School, and the material on Yangjuan is a combination of both of our experiences.
10. Interview with county official from the Yanyuan Education and Culture Office.
11. A developed township serves as a marketplace for surrounding villages' goods and as a link to the county seat. A market township is similar but provides few services beyond basic necessities. Developed townships may have other industries such as tourism. Markets towns are usually located in more remote areas.
12. Other ethnic minorities are primarily Zangzu (Prmi) and Mengguzu (Na). These local ethnic groups were assigned to the Tibetan and Mongolian *minzu* (nationalities) respectively as part of the Ethnic Identification (*minzu shibie*) processes in the 1950s and 1980s.
13. Interview with Shaba Elementary principal.
14. In 2007, the Baiwu principal reported that the school had seventeen positions for *minban* teachers, but had only been able to fill eight of these, because the low wages he could offer made the positions extremely unattractive.
15. The data do not account for possible differences in the age structure of the respective village populations. Such data are extremely difficult to find, especially at the village level, and often do not go back more than two years. The most precise measurements come from asking village-heads in person, who often must rely on memory. Nevertheless, the age structures would have to differ radically and improbably in order to account for the differences in ratios of school children to total population.
16. It is noteworthy, however, that in a larger-scale study of poverty and inequality among minorities in the Southwest, Bhalla and Qiu (2006) have claimed that, while location was a significant predictor of access to primary education among both Han and minorities in rural China in 1988, with children living in mountainous areas less likely to attend schools than those in hilly or plains areas, that effect had disappeared by 1995. It would be interesting to see if the effect would reappear nationally in statistics based on surveys taken after the 1999–2000 school consolidation (see Bhalla and Qiu 2006, 94–95, 98).

References

Bhalla, A. S., and Shufang Qiu. 2006. *Poverty and Inequality among Chinese Minorities*. London and New York: Routledge.

Bradley, David. 2001. "Language Policy for the Yi," in *Perspectives on the Yi of Southwest China*, ed. Stevan Harrell, 195–213. Berkeley and Los Angeles: University of California Press.

Hannum, Emily. 1999. "Political Change and the Urban-Rural Gap in Basic Education in China, 1949–1990," *Comparative Education Review* 43 (2): 193–211.

Hannum, Emily, Jere Behrman, Meiyuan Wang, and Jihong Liu. 2008. "Education in the ReformEra," in *China's Great Economic Transformations*, eds. Loren Brandt and Thomas Rawski, 215–249. New York and Cambridge: Cambridge University Press.

Harrell, Stevan, and Ayi Bamo 1998. "Combining Ethnic Heritage and National Unity: A Paradox of Nuosu (Yi) Language Textbooks in China," *Bulletin of Concerned Asian Scholars* 30 (2): 62–71.

Harrell, Stevan, and Erzi Ma. 1999. "Folk Theories of Success: Where Han Aren't Always the Best," in *China's National Minority Education*, ed. Gerard A. Postiglione, 213–241. New York and London: Falmer Press.

Harrell, Stevan, Qubumo Bamo, and Erzi Ma. 2000. *Mountain Patterns: The Survival of Nuosu Culture in China.* Seattle and London: University of Washington Press.

Mauger, Peter. 1983. "Changing Policy and Practice in Chinese Rural Education," *China Quarterly* 93: 138–148.

Park, Albert, Scott Rozelle, and Christine Wong. 1996. "Distributional Consequences of Reforming Local Public Finance in China," *China Quarterly* 147: 751–778.

Pepper, Suzanne. 1981. "Chinese Education after Mao: Two Steps Forward, Two Steps Back and Begin Again?" *China Quarterly* 81: 1–65.

———. 1990. *China's Education Reform in the 1980s: Policies, Issues, and Historical Perspectives.* Berkeley: Institute of East Asian Studies, University of California at Berkeley, Center for Chinese Studies.

Schoenhals, Martin. 2001. "Education and Ethnicity among the Liangshan Yi," in *Perspectives on the Yi of Southwest China*, ed. Stevan Harrell, 238–255. Berkeley: University of California Press.

Skinner, G. William, Mark Henderson, and Jianhua Yuan. 2000. "China's Fertility Transition Through Regional Space: Using GIS and Census Data for a Spatial Analysis of Historical Demography," *Social Science History* 24 (3): 613–652.

Tsang, Mun C. 1994. "Cost of Education in China: Issues of Resource Mobilization, Equality, Equity, and Efficiency," *Education Economics* 2 (3): 287–312.

Yanyuan Xian Zhi [Yanyuan Book of Statistics]. 1991. Yanyuan xian zhi bangongshi [Office of Yanyuan County Gazetteer].

Part III

Between Market Competitiveness and Cultural/Linguistic Identities

Chapter 9

The Relationship between the Trade Culture of a Hui Community and State Schooling: A Case Study of the Hui Community in Chaocheng, Shandong Province

Xiaoyi Ma

Chaocheng Township, Shenxian County, in Shandong Province has a dense Hui population. The Hui are Muslims and officially one of the fifty-five ethnic minorities of the People's Republic of China (PRC). As in other Hui communities across the country beginning in the 1980s, the market economy and *fumin* policy (a preferential policy that encourages the Hui to get rich) have had a great impact on Chaocheng's Hui population (see Gladney 1998). Relying on traditional industries such as leather processing, the local Hui people have rapidly improved their living and financial conditions over the past decade. Chaocheng Township overall has benefited from the revival of Hui small-scale industries and thus is economically better off than other townships in Shenxian County. However, the failure of the township's secondary schools to enroll and retain Hui students is a persistent problem—most of the secondary-school-age students in the local Hui community have dropped out of school. As a consequence, the educational level of the Hui community is lower than the county's average. Surprisingly, the rapid economic development of the community has had almost no effect on raising the township's educational level (see Ma 2004).

With the exception of their observance of the Islamic prohibition against pork, in everyday life the Hui people living in Chaocheng have much in common with the local Han people. There is little else that marks them as Hui, so they are not like the more typical Hui communities in northwestern China that preserve more traditional Islamic culture. In fact, research on Hui education, particularly on elementary education, mainly focuses on Ningxia Hui Autonomous Region or other Hui communities in the northwest in relation to Islamic religious education (see Cai 2006; Ding 1991; Ma 1991). However, very few researchers have paid attention to the situation of a Hui community such as the one in Chaocheng. It has better living conditions and is less affected by Islamic religious education, but has a higher rate of school dropouts, relative to Hui communities in the northwest. In the long run, I believe that the lower educational level will disadvantage the Hui community as China's development as a modern information society leaves them behind.

In this chapter I examine the problems of low academic achievement and secondary school dropout rates in the financially well-off Chaocheng Hui community and analyze them in terms of "cultural conflict theory" (Fordham and Ogbu 1986; Ogbu 1974, 1978). I also try to sort out other factors responsible for these problems.

Background Information and My Approach

Ethnic and Religious Culture of the Hui

As an ethnic group, the Hui is one of the fifty-five ethnic minorities in China. Its formation is the result of multiracial and multiethnic integration over a long period of time. There are different theories about the origins of the Hui, including one that posits a common origin for both Han and Hui (see Yang and Ding 2001). However, the most widely accepted theory is that the Hui nationality was formed in the Yuan Dynasty (1206–1368) from people immigrating from Persia, Arabia, and the Middle East (Gladney 1998; Lin 1984, 54–76). As an ethnic culture, the Hui not only have much in common with other cultures, but also demonstrate a much more integrated culture of sanctity and secularity than China's other Islamic ethnic groups. In the process of formation, the Hui culture absorbed the traditional Chinese culture and integrated it with Islamic culture. As a subculture of the Chinese culture, it was affected greatly by Confucian ideas. However, the Hui culture differs from other Chinese cultural systems mainly in its inner core based on Islam. The core of Hui culture is faith in Allah (see Yuting 2002).

The organization of the Hui community is unique also in production and social relations (Ma and Shen 2001). Called *jiaofang* in Chinese Muslim terminology, a Hui community is a separate regional religious unit that is formed of Muslims living around a mosque. A *jiaofang* is localized and delineates a group of people sharing a certain mode of production and social relations. Generally speaking, each *jiaofang* has its own special financial base, such as a handicraft industry or particular kind of trade, that sustains livelihood for a Hui community. Thus, the Hui community has its own culture and way of life, language, and, most important of all, a distinct ethnic identity.

The Hui community in Chaocheng considers itself a *jiaofang*. Indeed, the Chaocheng Hui have their own mosque where they can complete their necessary religious activities, such as weddings, funerals, and important Muslim festivals.

Trade Culture of the Hui in China and that of the Hui in Chaocheng

As early as the Tang (618–907) and Song (960–1279) Dynasties, the ancestors of the Hui actively engaged in trade along the famous "Silk Road," by land from Xi'an to Persia, and the "Flavor Road," from Malaysia to the Persian Gulf by sea. During the Yuan Dynasty, after Islam was introduced to China, the ancestors of the Hui began to flourish in China's markets and made great contributions to economic and cultural exchanges between China and Arab countries. Jewelry, gold and silver, spices, leather, and so on were the traditional commodities of Hui trade, so that the Hui were called *shi bao huihui* (good-at-jewelry-trade Hui). Some of China's famous inventions, such as paper-making technology, the compass, and gunpowder, were introduced into Europe by Hui traders. Meanwhile, they also introduced Western medicine, astronomy, mathematics, and architectural and engineering technology into China. During the Ming (1368–1644) and Qing (1616–1911) Dynasties, the arrival of more Muslim merchants, visitors, and religious specialists not only generated China's Islamic culture, but also gave rise to a florescence of Hui trading traditions. Islam advocates trade, and the Koran includes economic visions and principles that have greatly affected Hui economic life.

The trade culture of the Hui discussed in this chapter reflects the larger culture complex of the Hui. On one hand, similar to the larger complex, it is affected by Islam. On the other, it is affected by the local community culture, including traditional value concepts, value orientations, ethnic psychology, and way of life. So the trade culture of the local Hui community that I will discuss here is different from that of the historic Hui traders discussed earlier. Here in Chaocheng, the local Hui trade culture basically means individual or family ownership of a trade or particular business. It requires neither advanced education nor high-technology skills. In fact, the local Hui people earn a relatively high income with only an elementary or middle school education. The local Hui people are content with this way of life and have passed it from generation to generation within the community. In addition, the trade culture of the Chaocheng Hui community is different from modern high-tech businesses in a contemporary market economic system.

My Approach

This study is based on one month of fieldwork in 2003 in my hometown of Chaocheng. Using a questionnaire, I collected basic information about the age, ethnicity, and religion of 130 student in grades 7 and 8 in a middle school in Chaocheng. I selected these two grades because they had the highest dropout rate. Of the 130 students, 30 were Hui and 100 Han. My interviews were carried out in three schools (Chaocheng Central Elementary School, Xijie Minzu Elementary School, and Chaocheng Central Middle School) and the local Hui community. The interviews included middle school students, some of their parents, and their teachers, headmasters, and administrators, as well as local community leaders.

In this study, I adopted the culture conflict theory in my frame of analysis. Culture conflict theory aims to explain the reason for minority students' lower academic achievement as due not to the extent of the differences between a minority culture and the mainstream culture, but to differences in particular characteristics between a minority culture and the mainstream culture, resulting in culture conflict (Ogbu 1995, cited in Ha and Teng 2001, 58; Ogbu 1974, 1978). In other words, culture conflict theory acknowledges that minority students' study style, values, attitudes, and behavior formed in their families and communities may clash with campus culture based on mainstream culture. According to the culture conflict theory, this clash is the reason why minority students have lower academic achievement compared with mainstream students in school. While this theory directly accounts for some of the reasons for high dropout rates among Hui students in the last years of middle school and after grade 9, I also identify other factors indirectly related to cultural differences that have a negative impact on Hui student attendance at Chaocheng's secondary schools, such as the nature of the standardized education system, the local implementation of national policies, and bias on the part of both teachers and parents.

Current Situation of Education in Chaocheng Hui Community

Enrollment data obtained from the two Chaocheng elementary schools show that 100 percent of the school-aged children in the community were enrolled in the elementary schools, and that there were no elementary-grade dropouts between 1999 and 2002. The data also indicate that all the sixth graders moved up to secondary school between 1999 and 2002.

The middle school data between 1999 and 2002, as shown in table 9.1, reveal few dropouts (1.2 percent) during grade 7, but more (19 percent) during grade 8 and significantly more (40 percent) during grade 9. Between 1995 and 2002, the rate of middle school graduates moving up to high school remained between 18 and 20 percent. For example, in 2002, there were only 88 grade 9 students who passed the entrance examination and moved up to high schools and technical/vocational secondary schools. Of the 18–20 percent of students who successfully moved up to high school from 1997 to 2002, 6 of those percentage points comprise students who entered Shenxian's key high school (key schools are generally more competitive than other schools). Students who enrolled in non-key high schools represented 8 percentage points, and students who went to technical/vocational secondary schools made up 4.

In recent years, the dropout rate from secondary schools reached as high as 50 percent or more, significantly exceeding the national average dropout rate of 2 percent. Among the dropouts in Chaocheng, Hui students comprised the majority. For example, statistics from one Hui neighborhood

Table 9.1 Dropouts Chaocheng Central Middle Schools, 1999–2002

	Grade seven	Grade eight	Grade nine
Total number of students	800	650	480
Number of dropouts	10	150	320
Dropout rate (%)	1.2	19	40
Percentage moving up to higher school	N/A	N/A	18

Source: Statistics Provided by Chaocheng Central Middle School.

Table 9.2 Teenage (aged 13–18) dropouts in a Hui neighborhood in Chaocheng in 2003

	Middle school age (13–15)	High school age (16–18)	Total secondary school age (13–18)
Total	111	85	196
Students at school	51	42	93
Number of dropouts	60	43	103
Dropout rate (%)	54	50.5	52.5

Source: Chaocheng Township's permanent residence registration.

in Chaocheng show that among 196 Hui teenagers between ages 13 and 18, 103 (52.5 percent) dropped out of secondary schools or chose not to continue their secondary education (see table 9.2).

Among students who go on to high school, those who are academically well prepared and can afford it go to the county's key high schools or those in neighboring cities. These students usually go on to college. However, students who are academically less prepared or cannot afford to attend key schools end up at the local high school, where few graduates can typically pass the college entrance examination. Under these circumstances many academically underprepared students and their families decide that it is a waste of time and financial resources to continue their secondary education because they have no opportunity to go to college.

Theoretical Explanation of the Hui Trade Culture and State Schooling

Hui students living in an environment with strong business culture, such as in Chaocheng, usually participate in family trade activities, which deeply influence their value orientations and reduce their interest in schooling. In their value orientations, they usually see short-term financial interest as more important than continuing in school. In this kind of Hui community, influenced by the environment and peers, boys particularly tend to develop a narrower concept of employment, that is, considering only jobs in family trade and craft production. Because there is some conflict between formal education (including its value orientation, its subjects of learning, and its worldviews) and the Hui community culture (including its value orientation, way of life, and mindset), Hui students often resort to their community culture once they and their parents lose their interest and trust in formal education. Thus, in the short term, community trade culture develops a supportive environment, since it provides a job market or opportunities for secondary school dropouts.

As explained earlier, culture conflict theory claims that low academic achievement by minority students is not caused by the extent of the differences between minority culture and the mainstream culture, but by differences in particular characteristics between minority culture and the mainstream culture, resulting in cultural conflict (Ha and Teng 2001, 58). To some extent, it is reasonable and valuable in applying culture conflict theory to explain low academic achievement by Hui students from the perspective of family and community, though the difference in China is ethnic rather than racial.

In Chaocheng, state schools are the media transmitting the modern mainstream culture, but the Hui community and families are the carriers of the traditional Hui culture. State schools transmit knowledge of modern sciences and technology and are themselves cultural media that carry the framework of knowledge, worldview, and value orientations of the mainstream society. The Hui community and families intend to enable their children to inherit their traditional value orientations, attitudes, behaviors and ways of life. Thus, there are inevitable conflicts between the Hui community/families and state schools, as their cultures clash. In the Hui community in Chaocheng, the community culture mainly presents itself in the form of small-scale industries with simple technology and trade. In this environment, it is difficult for children to recognize the importance of receiving more education in public schools. This is particularly the case when Hui students do not feel academically successful in school; they will easily give up formal education for family business. Therefore, I think that higher secondary school dropout rates in the Hui community represent conflict between the mainstream school culture and the traditional Hui community culture, but conflict that is exacerbated by other factors as well. Below, I summarize the culture conflict and then discuss other conflicts that complicate and worsen the problem of cultural differences between the Hui community and state schooling.

First, there is a conflict between the public schools' educational values and those of the community. Public schools represent and spread the mainstream culture, that is, the Han culture. The mainstream culture features Confucian educational values that promote elite education, even in secondary schools, preparing students for college entrance examinations and for service in the government or state businesses. Though recently there has been a push to transition from this kind of elite education in order to nurture innovative and quality talents, it is difficult for reform to take place if national college entrance examinations are still given annually. In addition, the Hui community has its specific culture, characterized by a mosque that spreads the Islamic culture, including valuing family business as a strategy for making a living from generation to generation. These two different cultures share an emphasis on education, but also conflict over its function. Currently, formal education in Chaocheng is still elite education, mainly oriented toward success in the college admission examinations. However, Hui families value education as teaching practical skills necessary to the family business, With a means of livelihood at hand, Hui students and their families see few reasons for preparing for competitive exams to gain entrance to competitive schools. Thus, by the time of their completion of nine-year compulsory education, most Hui students have little opportunity for attending quality high school and higher education, which in turns limits their opportunities to become members of the mainstream society—an industrialized information society. When they choose not to continue their education, they choose Hui cultural values and way of life, though the choice is often a difficult one.

Second, there is a conflict between cultural diversity and China's modern education system. China is a multinational state and multicultural society with fifty-six nationalities. From the perspective of culture relativism (Zhuang 2001), each national culture has its own rationality. Based on China's constitutionally guaranteed national equality (see chapters one and two), each different national culture should be respected. Minorities are assured by the PRC Constitution to have equal opportunity to receive an education in a multinational culture. As a subculture of China, Hui culture is characterized by a trade culture that has affected the way of life for generations of Hui. However, China's modern education system, within the framework of the mainstream culture, requires that every school-aged child receive a nine-year compulsory education that does not accommodate the Hui culture. Thus, there is a temporary conflict leading to higher rates of secondary school dropouts. It is likely that the Hui trade culture will need the support and empowerment of a modern education in order to sustain itself in a well-developed market economy and modern society. Therefore, China's education system should improve to accommodate the nation's cultural diversity, while China's minorities need to accommodate modern education.

Third, there is a conflict between state educational policies and local implementation. The Hui community, as a part of the multinational China, must comply with state policies. Though the Hui community has its own traditional culture that promotes family business or trade among the youth, the Hui community also encourages school-aged children to receive the nine-year government-mandated education, which is considered the duty and obligation of every Chinese citizen. In the Hui community of Chaocheng, on the one hand, every family sends their children for the nine-year compulsory education while, on the other, more and more children drop out of schools toward the last phase of those nine years or beyond grade 9. The higher dropout rate appears to reflect a clash between state polices and a community's compliance, but in fact this is a compromise that helps to deflate the conflict: Hui children basically meet the state requirement for compulsory education, but very few of them go beyond it.

Finally, there is a conflict between school administration and students. In my fieldwork I found that the administrators and some teachers in the middle school in Chaocheng were prejudiced against academically underachieving Hui students. School administrators paid little attention to these students and took few measures to help them. Meanwhile, not receiving any warm encouragement from school, these students gradually lost their interest in learning, leading to poor academic achievement and even dropping out of school. The situation created and perpetuated a vicious circle. Modern schooling provides students with a lot of book knowledge, but the Hui community gives these students a direct and quick way to make a living. The two different cultures coexist in a limited sphere without enough accommodation, resulting in poor academic achievement and a higher dropout rate among Hui students.

Accommodating the Hui Culture in Formal Education

A healthy and benign interactive relationship should be constructed between the trade culture of the Hui community and formal education. The community and schools should communicate more often with each other about students' behavior and academic achievement in school. The Hui community should cooperate with the school administration and provide schools with feedback, such as how to deal with Hui students with behavioral problems and how to handle the relationship between parents and the school. It is essential for the community to help local schools understand the culture of the community and promote communication between Hui and Han students and teachers, so that cultural conflicts may be avoided and normal and effective teaching can be carried out in school.

I have stated early on that the Hui community has its own culture and way of life, and, most importantly, strong awareness of ethnic identity and mentality. Such awareness and mentality among children in Chaocheng's

Hui community may manifest as barriers to contact with Han students and teachers. When teachers do not handle Hui students' behavioral and academic problems properly, Hui students strongly and directly associate the situation with ethnic differences. Some Hui parents who consider such issues as ethnic discrimination go so far as to defend their children's undesirable behaviors at school. Here we should resolve two problems. First, Hui parents should develop more positive views of ethnic differences and overcome narrow-minded views when they confront general problems in school management, particularly problems concerning their own children. They should have more communication with their local schools regarding their children's social and academic growth. Second, the school administration also should pay more attention to ethnic differences and related problems in the classroom. For example, in Han-dominated secondary schools, not only should the school authorities pay special attention to these students and take care to show them what is academically and socially expected in school, they should also take measures in public relations and cultural awareness. The school could promote Hui history, customs, and culture, for instance, by providing information on these topics in school newsletters and on school public announcement boards. This kind of public drive can facilitate Han students' understanding of the Hui people and their culture. And more importantly, it will promote Hui students' ethnic pride. Such accommodating measures will not only cultivate a healthy learning environment for both Hui and Han students at school, but also encourage smooth interaction between school culture and community culture.

I also think that the government should more effectively enforce the nine-year compulsory education law to reduce the dropout rate of the Hui students. Schools, families, and the community should all pay more attention and give active guidance to the Hui students. Finally, state education should also be examined. There is too much emphasis on mainstream (Han) culture and educational context, while denying a place for minority cultures, and too much concern for national integration, while neglecting the importance of ethnic identity.

References

Cai, Guoying. 2006. *Zhongguo ningxia huizu jiaoyu* [Education for the Hui People in Ningxia Hui Autonomous Region in China]. Beijing: Kexue chubanshe.

Ding, Hong. 1991. "Huizu jiaoyu wenti guan kui" [Views on Problems in Education for the Hui People]. http://library.jgsu.edu.cn/jygl/gh02/LWJ/1952.htm [in Chinese].

Fordham, Signithia, and John Ogbu. 1986. "Black Students' School Success: Coping with the Burden of 'Acting White,'" *Urban Review* 18 (3): 176–206.

Gladney, Dru. 1998. "Getting Rich is Not So Glorious: Contrasting Perspectives on Prosperity among Muslims and Han in Deng's China," in *Market Cultures: Entrepreneurial Precedents and Ethical Dilemmas in East and SE Asia*, ed. Robert Hefner, 104–128. Boulder, CO: Westview Press.

Ha, Jingxiong, and Xing Teng. 2001. *Minzu jiaoyuxue tonglun* [An Introduction to Minority Education]. Beijing: Jiaoyu kexue chubanshe.

Lin, Gan. 1984. "Shi lun Huihui minzu de laiyuan yu xingcheng" [Discussion on the Origin and Formation of Hui Nationality], in *Huizu shi lunji* [A Collection of Hui History], 54–76. Yinchuan: Ningxia renmin chubanshe.

Ma, Enhui. 1991. "Shandong bufen huizu zuyuan wenti tansuo" [Discussion on the Origin of Parts of Hui Nationality in Shandong province], *Ningxia shehui kexue* 47 (4): 43–46.

Ma, Xiaoyi. 2004. "Huizu jingshang wenhua he xuexiao jiaoyu guanxi yanjiu" [Research on Community Trade Culture of the Hui and Schooling Education], master's thesis, Central University for Nationalities, Beijing.

Ma, Yinian, and Hui Shen. 2001. "Huizu yuyan, huizu jiaofang he huizu shequ you'er jiaoyu" [Language of Hui Nationality and Mosque Surrounding Areas and Kindergartens in Hui Nationality Community], *Minzu jiaoyu yanjiu* 45 (4): 22–25.

Ogbu, John. 1974. *The Next Generation: An Ethnography of Education in an Urban Neighborhood*. New York: Academic Press.

———. 1978. *Minority Education and Caste: The American System in Cross-Cultural Perspective*. New York: Academic Press.

Yang, Daqing, and Mingjun Ding. 2001. "20 nianlai huizuxue redian wenti yanjiu shuping" [Comments on Hot Issues in the Study of the Hui in the Last 20 Years], *Huizu yanjiu* 48 (4): 53–58.

Yuting. 2002. "Shi lun Huizu wenhua de shensheng xing yu shisu xing de liangxing hudong" [Discussion on Benign Interactivity of Sanctity and Secularity of the Hui], *Huizu yanjiu* 49 (1): 97–100.

Zhuang, Kongshao. 2001. "Jiazu yu rensheng—hubian yehua" [Clan and individual life—A Conversation by the Lakeside]. Wuhuan: Hubei jiaoyu chubanshe.

Chapter 10

Issues of Minority Education in Xinjiang, China

Rong Ma

Formal education in local minority languages in minority autonomous regions has been the official policy since the 1950s in the People's Republic of China (PRC; Xie 1989). It consists of a dual system extending from kindergarten to university, in which the "Chinese schools" use *Putonghua* (Mandarin) as the language of instruction and the "minority schools" use local minority languages as the media of instruction. In the minority school system, in general, Chinese language courses are offered as a second language, and all other courses are taught in minority languages. This comprises the bilingual education system in minority regions in China.[1]

The Xinjiang Uyghur Autonomous Region is the largest minority region in northwestern China, with an area of 1.65 million square kilometers. Based on research findings on minority education in Xinjiang during the period 2002–2007, this chapter discusses several key factors affecting minority education in Xinjiang today. This project was funded by the PRC's Ministry of Education and organized by Rong Ma from Beijing University in collaboration with scholars and graduate students at Xinjiang Normal University.[2] The research team visited government bureaus of labor and personnel, as well as universities and a number of secondary schools in Urumqi City, Tacheng, Aksu, and Kashgar Prefectures. Through the assistance of these institutions, over a hundred college/university graduates were interviewed.

Interviews with school principals and teachers, student graduates and parents, and government officials in labor management and personnel bureaus identified several key issues related to the current minority education system. The local interview data obtained in this project show the general patterns of formal education and employment after graduation. The information provides useful insights for a better understanding of the reality of minority education in Xinjiang, as well as in China in general.

Rapid Growth of Educational Development in Xinjiang since the 1980s

The policies and general status of minority education in China during the 1950s to the 1980s have been discussed in other chapters of this book. This chapter focuses on bilingual education in Xinjiang.

Table 10.1 shows a trend of increasing college admission in Xinjiang over the past twenty years. During 1977–97, the total number of college

Table 10.1 The growth of college admission in Xinjiang (1977–97)

	College admission				
Year	Total admission	To colleges in Xinjiang	To colleges in other provinces	Total applicants	Percentage enrolled
1977	3,916	2,938	978	109,577	3.6
1978	4,930	3,816	1,114	66,504	7.4
1979	4,266	3,224	1,041	54,728	7.7
1980	4,807	3,346	1,461	60,370	7.9
1981	4,409	3,063	1,346	67,114	6.6
1982	5,568	3,795	1,771	69,504	8.0
1983	7,761	4,967	2,794	77,621	10.0
1984	10,273	5,653	3,887	77,985	13.2
1985	12,000	6,741	5,259	69,673	17.2
1986	11,785	6,458	5,327	58,585	17.5
1987	12,939	7,281	5,658	73,947	17.5
1988	14,690	9,211	5,479	85,718	17.1
1989	13,405	6,903	6,502	86,499	15.5
1990	12,965	6,577	6,388	81,062	16.0
1991	12,791	6480	6,374	79,000	16.2
1992	17,069	9,318	7,751	84,518	20.2
1993	20,143	10,475	9,668	78,675	25.6
1994	17,839	9,000	8,839	60,119	29.6
1995	17,814	9,200	8,614	54,561	32.6
1996	17,737	9,260	8,477	55,360	32.0
1997	19,299	9,496	9,803	52,381	36.8

Source: University Admission Office of Xinjiang, 1997, 595–596.

student enrolled grew from about four thousand to about twenty thousand, a fivefold increase. About ten thousand of these students entered universities in Xinjiang; most of this group would seek employment in Xinjiang after their graduation.

In the late 1970s, only 7 percent of high school graduates were admitted to universities for higher education. However, by 1997, 35.6 percent of Chinese high school graduates and 40.5 percent of minority high school graduates went to college.

Table 10.2 shows school enrollment in Xinjiang between 1980 and 2005. In the later 1990s, there were about 5,000 college graduates and 10,000 two-year college graduates in Xinjiang every year. By 2005 there was a much larger college enrollment in Xinjiang—about 58,653 college students in total. In Xinjiang, one of the most serious social issues is that since the 1990s a large proportion of college graduates have been unable to find jobs. Due to the small number of employees who annually retire,

Table 10.2 New student enrollment by level and type of school, 1980–2005

		Secondary school						
Year	College[1]	Middle school	High school	Specialized secondary school	Vocational secondary school	Technical school	Primary school	Special school[2]
1980	3,767	23,6972	73,959	14,688	792	10,531	422,089	90
1985	9,298	247,431	93,692	13,919	20,429	10,603	374,643	91
1990	8,034	213,074	84,567	15,586	27,968	15,420	335,552	103
1991	8,179	161,986	77,525	16,902	29,354	16,722	344,432	15
1992	10,985	208,186	6,1673	20611	27,041	17,194	359,302	119
1993	13,359	207,503	56,180	26,572	24,778	16,851	385,535	102
1994	12,099	214,685	44,552	29,248	21,350	16,325	400,700	102
1995	12,307	228,115	58,203	25,260	25,643	19,593	419,717	165
1996	12,421	239,948	59,400	26,311	23,003	18,110	428,651	67
1997	12,673	262,117	64,377	27,302	21,741	17,034	443,238	120
1998	12,880	293,932	71,302	27,497	19,862	13,807	429,247	130
1999	19,821	329,715	69,801	36,014	18,828	9,221	391,298	175
2000	30,689	342,046	76,744	38,866	20,027	9,080	36,4193	177
2001	42,253	366,359	93,519	26,763	16,621	7,207	352,975	135
2002	42,808	388,558	113,366	22,241	16,746	10,442	346,386	874
2003	44,733	406,824	126,163	24,016	15,378	10,652	347,619	369
2004	53,204	398,090	138,315	27,990	27,585	14,686	347,364	390
2005	58,653	390,224	145,044	34,882	12,416	15,115	338,539	847

Notes: 1. College admission in this table includes two-year junior college enrollment. 2. These are schools for blind, deaf, and mute children.

Source: Statistical Bureau of Xinjiang, 2006, 518.

government institutions cannot create 58,000 new positions every year. The rapid expansion of college admission in the past several years has made the situation even worse. Meantime, the total number of government employees has stayed almost the same. Since the early 1990s, all college graduates have had to find jobs on their own after graduation. The difficulty in job hunting was the key issue in our survey interviews in Aksu and Tacheng Prefectures.

Table 10.3 gives us an overall picture of schooling, graduation, and the employment situation in Xinjiang between 1980 and 2005. In 1980, only 6.2 percent of high school graduates could receive college education. By 1997, 35.6 percent of Chinese high school graduates and 40.5 percent of minority high school graduates went on to college. The percentage rose again by 2002, when about 69 percent of all high school graduates went to college. This growth may have reduced the quality of college students, which would also have some negative impact on their job hunting after graduation.

Table 10.3 Student graduation and employment, 1980–2005

Year	Primary school graduates to middle school (%)	Middle school graduates	Middle school graduates: To high school (%)	To other schools (%)	To labor market (%)	High school graduates: To college (%)	To labor market (%)	College graduates
1980	81.0	186,012	39.8	14.0	46.2	6.2	93.8	814
1985	85.6	199,206	47.0	22.6	30.4	13.7	86.3	3,511
1990	82.2	216,173	39.1	27.3	33.6	9.9	90.1	8,603
1991	88.3	195,249	39.7	32.3	28.0	10.3	89.7	7,919
1992	77.7	176,435	35.0	36.7	28.3	13.0	87.0	8,402
1993	80.0	182,827	30.7	37.3	32.0	17.0	83.0	7,741
1994	81.9	141,749	31.4	47.2	21.4	21.2	78.8	7,734
1995	84.1	171,167	34.0	41.2	24.8	27.6	72.4	10,505
1996	85.7	171,817	34.6	39.2	26.2	28.6	71.4	12,272
1997	89.0	190,321	33.8	34.7	31.5	33.4	66.6	10,908
1998	91.6	210,208	33.9	29.1	37.0	25.9	74.1	11,401
1999	92.2	230,083	30.3	27.8	41.9	37.8	62.2	118,86
2000	92.0	253,240	30.3	26.8	42.9	55.8	44.2	10,985
2001	92.7	270,550	34.6	18.7	46.7	71.3	28.7	16,121
2002	94.4	294,271	38.5	16.8	44.7	69.1	30.9	16,380
2003	96.5	308,445	40.9	16.2	42.9	63.2	36.8	25,785
2004	95.3	336,165	41.1	20.9	38.0	63.0	37.0	31,013
2005	98.1	356,359	40.7	17.5	41.8	56.4	43.6	37,920

Source: Calculated from the data published by Statistical Bureau of Xinjiang, 2006, 518–519.

The Market Mechanism in Employment of Various Levels of School Graduates

Transition from Planned Economy to Market Economy

Minority education has always faced many issues, such as a shortage of qualified minority teachers, a dearth of quality textbooks containing local knowledge of ethnic minorities, few connections between minority schools and Chinese schools, and a lack of minority language teaching facilities in some disciplines. However, some new problems have emerged since the 1980s. China has experienced a tremendous social transition in the past thirty years, including a series of reforms in the agricultural community structure (e.g., the disintegration of communes) and in the industrial and trade sectors (e.g., the transformation of state-owned enterprises into diversified ownership systems). China is still transitioning from a centrally planned system into a market-oriented system. These changes certainly have had some impact on China's minority education system.

Under the planned-economy system, until the late 1990s, all graduates of universities and vocational secondary schools (three-year training programs after middle school) were guaranteed jobs as employees of the government, its institutions, and businesses. At that time, all universities, schools, research institutes, publishing houses, theatrical companies, film companies, hospitals, factories, transportation companies, and so on were under government management and recruited their employees according to government plans. All employees in those organizations received salaries and benefits from the government. In the planned system, ethnic minority graduates usually were assigned jobs as government officials because of shortages of educated minority cadres. Under the planned-economy system, state-owned institutions did not need to worry about efficiency and financial resources for their employees. And sometimes, the recruitment of minority graduates was considered a political task, the implementation of the government's affirmative action policies for ethnic minorities.

After the reforms, the situation changed completely. Under the new policies during the transition from the planned economy to a market economy, state-owned enterprises were transformed into private enterprises, joint ventures, or public companies (selling shares on the market). And these reformed enterprises follow market practices as the main mechanism in their management and employment. Now enterprises recruit and dismiss employees at their own discretion and employees no longer have lifelong jobs. Enterprises hire or discharge their employees according to their own needs and the ability and skills of the individuals involved, regardless of their ethnicity. Minority students who are admitted to college under affirmative action policies are likely to be confronted with employment difficulties if they are insufficiently competitive in their professional ability upon graduation. Even if they are recruited, there is still a possibility they will be fired if they do not pass competency evaluations. Therefore, minority students face intense competition in the new, non-state-owned labor market. Some of them have substantial difficulties finding and keeping jobs due to poor professional preparedness, working skills, and *Putonghua* language proficiency.

Reduced Government Employee Hiring Quotas

Since the late 1980s, the Chinese government has released many of its administrative functions to social organizations, such as companies or associations, especially in the economic sector. Many economic activities are now managed by the market, not by the government. Thus, administrative employment has been reduced at all levels of government. In Aksu Prefecture, for example, government institutions at the prefecture level were requested to reduce their employees by 23 percent, at the county level by 18 percent, and at the township level by 13 percent.

As a result, reduced hiring quotas at government institutions have spawned complaints from both government labor bureau officials and students. In China, public school teachers and doctors and nurses in public hospitals and clinics are government employees. Institutions such as schools and hospitals must have a new quota from the government personnel budget in order to recruit any new employee except for those filling retirement vacancies. These institutions' budgets cover salaries and other benefits (health care, housing, etc.) only for employees counted in their quota.

Many minority students who graduated from Xinjiang Normal University have returned to Aksu and Tacheng to seek jobs since the late 1990s. The schools and hospitals they have contacted wanted to recruit them, but they could not obtain quotas from the government. The reason given by the prefecture personnel bureaus was that the total of their present quotas exceeded the size in the available government budget. For example, in 2003, the First Secondary School of Aksu officially had 166 positions (as per quota), but it actually had 200 employees on its payroll. In the school system in Aksu Prefecture, there were 27,000 teachers and 480,000 students, about 1 teacher per 17.8 students, which is a very good ratio. In other regions in China, it is usually about 1 teacher per 25–30 students. Therefore, the schools cannot hire new graduates because they have no convincing reason to request more teachers.

Four other factors have led to the difficult situation for college graduates. First, ongoing reform of the government has continually required reducing the number of current administrative positions. As a consequence, in order to keep their employees in the face of reduced quotas, many government bureaus/institutions, such as educational bureaus or public health care bureaus, have arranged to "borrow" the extra employee quotas from schools and hospitals under their management. Thus, these schools and hospitals actually have lost their quotas to their supervising institutions. The "extra" officials may still work in education bureaus, but they are registered as employees in schools. This means that there are fewer teachers and medical personnel on the ground than on the books. It is a common practice in Xinjiang.

Second, many minority students still consider only a government position to be "a decent" job and prefer to wait for a position in urban areas, refusing to work in small, private businesses or become self-employed. Because in the past (the 1950s–1980s) the small number of minority college graduates meant that they all became minority cadres in the government, both college students and their parents often still expect future government employment for graduates. This is a common phenomenon found in all minority regions, including Xinjiang, Inner Mongolia, and Tibet, but the situation started to change when the enrollment of minority students in universities rapidly increased in the 1990s. From our interviews, we learned that some college graduates finally started to seek employment in the private job market only after a series of failures to obtain a government position.

Third, a decline in the birth rate has reduced the need for teachers because there are fewer students attending schools today as compared with the 1980s.

Fourth, because more minority students than before are attending Chinese schools, the student population in minority schools has been further reduced. Some parents believe that learning Chinese and taking courses (mathematics, physics, etc.) taught in *Putonghua* will help their children to find jobs in the future. In Tacheng, minority students in large numbers have recently flocked to Chinese schools (about 70–80 percent in urban areas), reducing enrollment in minority schools and resulting in a surplus of teachers in those schools.

Corruption is another problem. When there are quotas and intense competition for scarce resources, corruption occurs. Many interview respondents complained that the children of local officials got jobs in government institutions while the children of ordinary citizens had been waiting for employment for years.

Another result of economic reform is that almost all state-owned enterprises have been bankrupted or transformed into private enterprises.

During the planned-economy period, those state-run enterprises were the main employers for college graduates, especially vocational school graduates. This change has had a crucial impact on employment for college graduates.

Discrimination against Minority Students in the Labor Market

Almost all minority college graduates whom we interviewed complained about ethnic discrimination in the labor market. The government has clear regulations prohibiting "ethnic preference" in recruitment, but according to our respondents, such "preference," or discrimination, in fact exists for three groups of students. In general, it is easiest for the group of college graduates called *Han Kao Han* (Han, Hui, and Manchu graduates of Chinese schools who took the college admission examination in Chinese) to find employment. The second group, known as *Min Kao Han* (minority graduates of Chinese schools who took the college admission examination in Chinese), also have an employment advantage. The job search is most difficult for the group of college students known as *Min Kao Min* (minority graduates of minority schools who took the college admission examination in minority languages).

The director of the Center of Personnel Exchange in Aksu Prefecture reported that there were about 5,650 graduates registered with his Center seeking jobs in 2003. Among them, 65 percent were females, 88 percent were ethnic minorities, and 61 percent were graduates from vocational schools; 38.6 percent were graduates from two-year colleges, and only 7 percent were four-year college graduates.

The main reason that minority graduates are discriminated against in job markets is their poor *Putonghua* proficiency. Most employers in the job market are Han managers representing local enterprises or those from eastern provinces. The employees and customers of their businesses are mainly *Putonghua* speakers. Many economic activities actually are based in coastal regions but are penetrating into western regions (Xinjiang). The function of language as a communicative tool plays a clear role in this circumstance. In general, this practice involves little cultural prejudice or discrimination but is linked to concerns for convenience and efficiency in business. In our local interviews, a Uygur respondent in Tacheng told us that even a Uygur businesswoman refused to hire her because she could not speak fluent *Putonghua*. There is another example from our interviews. In the first year of his business, the largest local Kazak private business owner hired his

Kazak relatives and friends. He almost went into bankruptcy within three months because his relatives were hard to manage. In the second year, he hired 50 percent Uygur employees and 50 percent Han employees. In the third year, only three Uygurs remained and over 80 percent of his employees were Han.

A common language is a necessary communication tool. Minority education faces an obstacle if minority students cannot properly master in school the common language. The language barrier essentially handicaps minority students in the job market. However, there is also cultural and ethnic prejudice in employment. Based on our field interviews, additional reasons why employers or managers did not want to hire minority graduates can be summarized as follows.

Because of ethnic tensions in Xinjiang, managers worry that minority employees might make ethnic relations in their companies more complicated. Furthermore, managers consider minority graduates' Muslim customs (diet and Ramadan) and mentality particularly inconvenient for management and a source of potential conflicts among employees. In addition, the affirmative action school admission policies for minority students give managers the impression that the abilities and skills of minority graduates might be lower than those of Han graduates. Managers also cite the poor performance of their past minority hires and the need for re-training programs for minority students because of their low competence in *Putonghua*.

Disadvantages for Minority Students in the Job Market

Minority Students' Putonghua Proficiency

Often, minority graduates who took Chinese language courses in minority schools did not have the chance to practice it, because often their Chinese language teachers were minorities and could not speak fluent *Putonghua* themselves. For example, there were only four Han teachers among a total of twenty-three who were teaching Chinese courses at Aksu Normal School in 2003. We were told that, unfortunately, these Han teachers were applying for transfer to other regions.

After the Cultural Revolution (1966–76), so many Han teachers working in minority schools transferred to other regions that Chinese language teaching stopped in Xinjiang for a period of time, according to a 1983 Xinjiang Education Commission document. Only in 1987 did the Xinjiang government start to emphasize that "Chinese is the major language of China. It is the major tool for all groups to communicate with each other in sciences and cultural exchanges. Learning and using *Putonghua* is needed for developing material and spiritual civilization construction" (Minority Education Bureau 1995, 337 and 808). Although minority schools in Xinjiang resumed teaching Chinese after 1987, the period of 1976–87 has left a negative legacy for Chinese language teaching in Xinjiang.

A vice-principal of Aksu First Secondary School told us that current minority students were eager to learn Chinese, but they were not happy with the proficiency of their Chinese language teachers who were native speakers of Uygur. Even if these teachers passed Chinese language exams in Urumqi, their pronunciation changed after they had spent four or five years back in their native community in Aksu. Some minority teachers, who graduated with a major in Chinese from Xinjiang University in 2000 and passed the grade 6 level of the Chinese proficiency examination (HSK), were still not qualified to teach Chinese in this school. A government official of the Aksu Bureau of Cultural Affairs mentioned that some minority college graduates who passed grade 8 level of HSK could not write an application or a report in Chinese.

In the urban school system in Tacheng Prefecture, the number of minority teachers exceeded the quota by 240 while there was a shortage of about 120 Han teachers. Some teachers had to teach courses that were not in their field of training. This situation reduced the quality of the courses. The poor quality of courses in the Chinese language and the sciences in minority schools drew many criticisms from parents and the local community. According to a teacher at the Aksu First Secondary School, many parents in the past did not want to send their children to study in the special "Xinjiang Class" in eastern (Han-dominant) provinces, but after 2003 about 30–40 percent of his students wanted to go to China's eastern provinces to receive a better education for their future careers.

Minority Students' Low Examination Scores

Another reason for minority graduates having difficulties in finding employment is the general impression that minority students are underprepared. This impression is directly related to the different college and secondary school admission standards for Han students and minority students.

A vice-principal of Aksu First Secondary School estimated that the average difference in the admission examination score between Han students and minority students in his school was around 200 points. The principal of Aksu Normal School told us that for admission to this teachers' school, the standard for Han applicants was 320 points, and for minority applicants it was 210–220 points.

According to unpublished records provided by the Education Commission of Xinjiang, the percentage of Han students in middle schools in Xinjiang who passed the exams for three science courses (physics, mathematics, chemistry) was 50.9–75.1 percent between 1994 and 1996, while the comparable percentage for minority students was 11.2–28.8 percent for the same period (Li 2003, 15). In 1996, 1997, and 1998, the percentages of minority students whose exam scores were above sixty was 11.2, 12.4, and 21, respectively, for mathematics; 23.3, 12.7, and 37.1 for physics; and 28.8, 29.4, and 42 for chemistry. It should be noted that in those three years, the percentage of minority students who could pass Chinese exams was 19.5, 29, and 26.1, respectively (p. 15). These data clearly show minority students' academic status in middle schools in Xinjiang.

Since minority students received low examination scores in their middle school courses, it is not surprising that the government had to lower the admission standards for them to enter high school. Table 10.4 shows the required score levels for high school admission in Urumqi City in 1991 and 2001. Since the contents and degree of difficulty of the exams in Chinese and minority languages were different, we can only compare Group I and Group II who took the same exams in Chinese. Our comparison finds that studying in the same Chinese school and taking the same exams made it much easier for the minority students to enter high school than their Han classmates (260 points versus 620). It should be noted that the gap was declining during the 1991–2001 period, mainly due to the rapid increase

Table 10.4 High school admission standards (exam scores) in Urumqi, 1991 and 2000

	Group I		Group II					
Score	**Han**	**Hui**	**Min Kao Han**		**Uygur school**		**Kazak school**	
1991	620	610	260		290		280	
	Han/Hui		**Min Kao Han**					
	A	B	A	B	A	B	A	B
2001	415	340	320	300	360	330	300	280

Notes:
Group I: Han and Hui students who are both native speakers of Chinese from Chinese schools; Group II: Minority students from Chinese schools.
A: Students who met the standards for the planned quota with regular tuition; B: Students who met the standards for the outside planned-quota with additional tuition.

Source: Huang 1997, 228; Li 2003, 16.

in the size of college enrollment in recent years. Some Western literature has analyzed the size of student enrollment but ignored the issue of different admission standards for minority students (Benson 2004, 196–205).

College Affirmative Action Admission Policy for Minority Students

The gap is obvious in college admissions for different groups of students. Table 10.5 presents the different standards in college admissions for Han and minority students in Xinjiang in 2002. The gap is larger in the fields of sciences and engineering than in humanities and social sciences. The low scores required for college admission for minority students in turn had negative impacts on their learning achievement in university and their ability upon graduation.

Table 10.6 presents college admission standards for different groups of students during 1977–2006. We need to compare only two columns: (1) scores for Han students from Chinese school, and (2) scores for minority students from minority schools. The year 1977 was a very special year: it was the time of the first national college admission exam after the Cultural Revolution. The gap between college admission standards for Han students and minority students who took the same exams for natural science disciplines was 170 points (300 versus 130) in 1980, 154 points (414 versus 260) in 1997, and 90 points (390 versus 300) in 2003. The gap was very significant and predictive of these students' academic performance in college. In 2003, with a lowest score of 18 points (total 100–150 points) for one of the

Table 10.5 College admission standards (national exam scores*) in Xinjiang, 2002

	Humanities and social sciences			Sciences and engineering		
Candidates	Top university	General university	2-Year college	Top university	General university	2-Year college
Group I	490	436	340	499	420	330
Group II	456	398	270	400	340	200
Group III	330	296	255	315	265	220
Group III**	16	16	12	18	16	12

Notes: *The national university admission examination includes five subject exams with a total score of 750.

Group I: Han students from Chinese schools; Group II: Minority students from Chinese schools; Group III: Minority students from minority schools; Group III**: The lowest score level of one of the five exams.

five subject exams of the national college admission exam, a minority student would still be able to go to college if his or her total score was above 265 points (out of the total of 750 points). Generally, minority students got their lowest scores in natural science exams, as shown in table 10.6.

There was no comparison for the minority and Han students who took different exams in different languages. But the quota system did guarantee a high percentage of these minority students entrance to universities. From table 10.4, we can see that a minority student could enter universities in the humanities and social science disciplines with a score of 345 (the required standard for the mathematic exam was 24 points out of a total 150 points) in 2005. In the same year, Han students could enter universities in the same disciplines only if they scored 380 or above. The gap was greatly reduced compared with the situation in 2002, when Han students were required to score 436 and minority students, 296. It should also be pointed out that the content of college admission exams for minority students has gradually followed that of the exams for Han students in order to increase the comparability.

We can look at more details about the gap between Han and minority college students. In 1998, the universities in Xinjiang recruited 4,248 minority students majoring in sciences and engineering. Of these students only 1.4 percent had a score above 60 (the passing level for the exams) in one of three subject exams (mathematics, physics, and chemistry), 60.2 percent had a score of 10–29, and 7.5 percent scored in the 0–9 range (Ren 2003). Clearly, these minority students would encounter difficulties learning, and their teachers, difficulties teaching. Even if they studied very hard, students' progress would be limited due to their poor academic foundation from high school.

The affirmative action policies for college admission also lead to corruption and cheating. Some Han or Hui students have tried to change their "ethnic status" to Uygur or Kazak in order to make their college admission easier.[3]

Consequences of the Affirmative Action Policies

According to the director of the teaching division at Aksu First Secondary School, reduction of college admission standards is the key factor responsible for the decline of educational quality in both secondary schools and universities. Many students who fail to reach the admission standards still receive a college admission letter and go to college. He cited the two-year program of the Law School and Adult Education School of Xinjiang University as examples of institutions that recruited low score students from his school. One of his students whose score was only 212 (the already lowered standard was 255 points) received an admission letter from the

Table 10.6 College admission standards (national exam scores*) in Xinjiang (1977–2006)

	Han (Chinese schools)						Minority (Minority schools)						M with C		Sports & art	
	H&SS			NS			H&SS			NS			H&SS	NS	Cul.	F.
Year	T	G	S	T	G	S	T	G	S	T	G	S				
1977	A65,	B80,	C65	A55,	B70,	C55	A55,	B65,	C55	A35,	B45,	C30	40	25		
1978		261			250			95			90		140	90		
1979		256			232			170			152		140	90		
1980	330	277		360	300		256	267		150	130	150	150	130		
								Mon 220			Mon 130					
1981		320			340			320			310		190	215		
1982	365	349		375	336			320/Mon320			367/Mon310		170	150	180	60
1983	425	412		440	405		425	400		500	460		200	180		
								Mon 300			Mon 390					
1984	428	415		405	365		320	297		350	300		180	155	200	70
								Mon 171			Mon 172					
1985	430	415	405	440	400	385	317	297	290	288	260	252	205	116	210	70
								Mon 115			Mon 170					
1986	450	440	430	470	450	425	245	235		335	300	285	195	210	280	75
								Mon 190			Mon 180					
1987	445	428	419	470	435	421	269	262	245	313	282	268				
								Mon 213			Mon242					
1988	463	453	441	480	453	443	305	304	290	423	385	370				
1989	467	452	443	498	474	453		322	308		364	347				

1990	434	418	409	500	473	454		247	264		282	277				60
1991	456	444	434	499	473	454		258	250		260	247				65
1992	458	445	435	523	500	473		283	276		285	275				
1993	440	430	420	482	458	443	320	280	273	322	287	282				
1994	488	471	454	522	485	446	410	368	342	378	335	312	343	340	343	446
1995	489	475	468	502	465	445	372	337	288	347	320	280	359	293	359	401
1996	526	516	481	498	446	424	375	322	—	326	—	402	326	231	320	
1997	478	460	444	468	414	388	371	335	320	332	288	279	326	260	326	
1998	464	434	404	484	432	404	343	324	309	337	309	301				
2000	464	434	390	478	422	388	330	304	285	324	290	260	376	300		
2001	468	436	344	486	436	344	—	—	—	—	—	—	—	—		
2002	490	436	340	499	420	330	330	296	255	315	265	220	398	340		
	(lowest score in any one of 5 exams)→						16	16	12	18	16	12				
2003	493	410	348	456	390	302	333	288	251	307	265	210	400	320		
	(lowest score in any one of 5 exams)→						18	18	10	18	18	10				
2004	538	437	320	522	397	300	381	345	297	382	352	297	399	373		
	(lowest score in mathematics)→						23	23	20	23	23	20				
2005	516	360	290	507	350	270	393	300	280	367	300	260	380	342		
	(lowest score in mathematics)→						25	23	22	27	25	24				
2006	517	380	295	520	370	275	398	345	290	357	310	270	440	382		
	(lowest score in mathematics)→						26	24	23	29	26	25				

Notes:

M with C: Minority students from Chinese schools; H&SS: Humanities and social sciences; Cul.: Cultural exams; NS: Natural sciences; F: Special field in sports and art; T: Top universities in Xinjiang; G: General universities in Xinjiang; S: 2-year college in Xinjiang.

1977, A: High school graduates in 1977; B: High school graduates in 1966 and 1967; C: Others;. Mon xxx: Admission standards for Mongolian students who take exams in Mongolian language.

Sources: Data for 1977–97: University Admission Office of Xinjiang, 1997, 597–598; for 1998–2006: Ma (2008), 38–40, cited from Li Xiaoxia (unpublished paper).

Adult Education College of Urumqi in 2002. Some universities have reduced admission standards to recruit more students just for their tuition. The total size of the student population has increased 50–70 percent at some universities in recent years, and this has further reduced the quality of their graduates. Meantime, many government institutions have organized their own "training programs." For example, with the cooperation of a local vocational school, the Prefecture Personnel Bureau in Tacheng has recruited students and offered degree certificates. The graduates from these programs also compete with other college graduates for jobs, and this worsens the unemployment problems for college graduates.

In our interviews, most minority teachers and students worried about the negative impact of the affirmative action policies for minority students overall in school admission. They believed that such policies did not motivate students to study harder, but increased the academic gap between Han and minority students and created the strong impression in society that minority graduates were not as qualified as Han graduates even if in fact they were excellent.

Minority College Student Subject Bias

A large proportion of minority students study humanities in college and concentrate in disciplines in their own languages, history, and philosophy. The job market can provide only limited opportunities for these majors. In addition, even the quality of programs in sciences and engineering for minority students is also reported as problematic. A graduate with a major in computer science from Xinjiang University reported that he mainly learned the principles of computing and had very limited practice in school. He said that he learned to use some useful software by himself after graduation.

Minority student bias in selecting college majors has some negative implications for standards in minority secondary education because many of these college graduates become secondary school teachers. This bias has resulted in a lack of qualified minority teachers in natural sciences, which in turn is not conducive to promoting interest in natural sciences among upcoming minority students in secondary schools. Our respondents referred to this as a "vicious circle."

Minority Students' Voluntary Choice of Language of Instruction

In order to promote employment for college and professional school graduates, the Xinjiang Autonomous Region government adjusted its language policy in 2004. Recognizing low *Putonghua* proficiency as one of the major obstacles for minority students in securing employment, the government now requires that courses in natural sciences in college be taught in Chinese. We discussed this new policy adjustment with minority teachers and students during our survey, and several issues emerged from our discussions.

First, minority students should have the right to decide which language they prefer as the medium of instruction in school. This is their legal right protected by the PRC constitution (see chapter two). It is questionable to enforce a language of instruction administratively and deny students their right to language choice. Second, many minority faculty members who used to teach mathematics or other science courses in minority languages now have to teach these courses in Chinese in order to keep their jobs, and their Chinese may not be good enough. This reduces the quality of their courses. Third, there are considerable difficulties for students from minority high schools in taking courses in Chinese in college. This will also reduce the quality of their college education in the short term.

The response of minority communities to the new policy for the medium of instruction was very negative in the first year. However, we learned that people started to accept it in the following years. It is believed that the employment pressure makes people face the reality of the world after graduation, and they begin to recognize that the new policy may have some positive outcomes. In addition, it should be pointed out that in the Marxist/Leninist tradition, the use of administrative measures to enforce Russian as the "official language" was criticized by Lenin in the 1920s, who warned that such measures would lead to opposite results. He believed that the development of market and trade would promote the use of Russian language spontaneously (Lenin 1913).

In our previous surveys, we learned that some local governments took the opposite track, insisting that all minority students must attend minority schools and receive their education in their mother tongue, as was the case in Tibet and Yanbian Korean Autonomous Prefecture (Liu 1989). For example, Lhasa Second Primary School recruited both Han and Tibetan students, but they were separated into Han and Tibetan classes using different languages of instruction. Its principal told us that this policy started to be enforced in 1988. Even if some parents asked their children to study in a "Han class," nevertheless, they were not allowed because it was the government policy to protect "minority languages and cultures."

There are also strong political pressures from the West, which considers Tibetan traditional language and culture to be in danger. Thus, this policy in Tibet is in part a response intended to deflect Western criticism. One interesting phenomenon is that some senior Tibetan officials have emphasized this policy at government meetings while sending their own children to study in Chinese schools in Chengdu, Sichuan Province (Ma 1996, 386). A similar phenomenon was also found in Yanbian, where the Han officials, rather than Korean officials, have strongly supported the policy to limit all Korean students to Korean schools because they are afraid of being accused of practicing "Han chauvinism" (Ma and Lamontagne 1999). To evade this policy, a Korean vice-president of Yanbian University sent all three of his children to study in Chinese schools in a nearby county, even though he had to pay an additional "transfer and non-residence fee."

Discussion

Language has two basic functions. It functions as the essential carrier of the cultural heritage and history of an ethnic group. It is also a tool enabling anyone to learn, via the mother tongue, from his/her own group, or from other groups by learning other languages. The legal rights of minority students to study in their mother tongues should be protected while the request of some individual members of minority groups to study in majority schools should also be respected.

> It is obvious that where speakers are part of a small language group and know only that language then their linguistic repertoire constrains them to social commerce within that group, often limits their choice of marriage partner and usually dictates where they can work. If the group's concern to maintain solidarity and continuity placed a limitation on the other languages that members may learn, this would be an infringement of individual rights. (Wright 2004, 226)

Regardless of whether the restriction in school choice comes from minority leaders, majority cadres, or even pressures from outside, it violates the basic right of an individual to select a language of instruction.

If minority students and their parents in China have the right to choose schools, some of them truly concerned about group identity, traditional culture, and group history will choose to attend minority schools, while others concerned about language as a communication and learning tool will choose Chinese schools to ensure future employment and opportunities. The two functions of language then will each have their proponents in our society. When most people receive what they desire, if their desire is proper, the society will be more harmonious than otherwise.

Language of instruction is a key and sensitive issue in minority education. Minority education as a system is a model to marry the interests of both ethnic minorities' concerns about their cultural heritage and their desire for national development and integration. Both levels of a "pluralist-unity" pattern should receive equal attention and be kept in balance.

Based on the information from field interviews in Xinjiang, it seems that the problems in the dual school system there are quite serious. The dual school system, with its different college admission standards, has negative impacts on both *Putonghua* proficiency and academic achievement among minority students, while both are critical factors in the job market after graduation. As the transition from a planned economy to a market-oriented economy has reduced government employment, and universities have increased the size of their enrollments, the two factors have worked in concert to exacerbate the situation. Universities need to adjust their teaching subjects and content in order to meet the needs of the labor market, which changes rapidly with economic transition and development. By encouraging Chinese teaching in schools and universities, the Xinjiang administration has taken the right course, but this radical action might cause some ethnic tensions. From the interviews with many minority teachers, students, and unemployed college graduates, we can feel their confusion, anxiety, and frustrated expectations for the future. As indicated earlier, currently many minority students criticize the affirmative action policy that resulted in their admission to college; they think they would have studied harder and achieved a higher level of learning, otherwise. Both the dual school system and affirmative action policies are complicated issues in the short and long term. They may be helpful to minority groups in the short term but harmful in the long run.

Since there are great variations among the ethnic minorities regarding their language traditions and their desire to maintain their traditional language, more research should be carried out in Xinjiang and other minority regions in China to study the problems and search for more alternatives for minority educational development.

Notes

1. There are some variations by region. For example, there are two types of bilingual teaching models in Yi areas in Sichuan (Teng 2001, 50–51). In some areas, Chinese is used to teach science courses such as mathematics, physics, chemistry, and biology.
2. Professors Cui Yanhu, Dimulati, Bahar and their three graduate students from Xinjiang Normal University and Ms. Li Xiaoxia from Xinjiang Academy of Social Sciences conducted the field interviews in Urumqi, Tacheng, and Aksu.
3. The Xinjiang government issued a special document in March 2003 to control the cheating activities and since then the government has enforced serious

punishment for students and other responsible personnel who engaged in illegal ethnic identity change for college admission.

References

Benson, Linda. 2004. "Education and Social Mobility Among Minority Populations in Xinjiang," in *Xinjiang: China's Muslim Borderland*, ed. S. Frederick Starr, 190–215. New York: M.E. Sharpe.

Huang, Jiaqing. 1997. *Xinjiang minzu xuexiao jiaoyu yanjiu* [Studies of Minority School Education in Xinjiang]. Urumqi: Xinjiang renmin chubanshe.

Lenin, Vladimir Il'ich. 1913. "Gei Shao Wu Mian de xin" [A Letter Addressed to Shaowumian], in *Liening lun minzu wenti* [Selected Writings of Lenin on National Questions], 253–256. Beijing: Minzu chubanshe.

Li, Xiaoxia. 2003. "Xinjiang gaoxiao zhaosheng zhong dui shaoshu minzu kaosheng youhui zhengce de fenxi" [An Analysis of Affirmative Action Policies for Minority Students in College Admission in Xinjiang]. Unpublished research report.

Liu, Qinghui. 1989. "Xizang jichu jiaoyu yu Zangyuwen jioaxue" [Basic Education in Tibet and Tibetan Teaching], in *Minzu jiaoyu gaige yu tansuo* [Reform and Exploration of Minority Education]. Beijing: Zhongyang minzu xueyuan chubanshe.

Ma, Rong. 1996. *Xizang de renkou yu shehui* [Population and Society of Tibet]. Beijing: Tongxin chubanshe.

——. 2008. "Xinjiang minzu jiaoyu de fazhan yu shuangyu jiaoyu de shijian" [Development of Minority Education and Practice of Bilingual Teaching in Xinjiang], *Beijing daxue jiaoyu pinglun* 6 (2): 2–41.

Ma, Rong, and J. Lamontagne (eds.). 1999. *Zhongguo nongcun jiaoyu fazhan de quyu chayi: 24 xian diaocha* [Regional Variation of Rural Education in Contemporary China: 24-County Survey]. Fuzhou: Fujian jiaoyu chubanshe.

Minority Education Bureau of State Commission for Ethnic Affairs. 1995. *Shengshizizhiqu shaoshu minzu jiaoyu gongzuo wenjian xuanbian (1977–1990)* [Selected Provincial and Autonomous Regional Documents on Minority Education, 1977–1990]. Chengdu: Sichuan minzu chubanshe.

Ren, Xinli. 2003. "Guanyu yizhi zongjiao dui jiaoyu shentou de duice yanjiu" [A Study of Controlling the Impact of Religion in Education], *Xinjiang shifan daxue xuebao* 3: 23–30.

Teng, Xing. 2001. *Wenhua bianqian yu shuangyu jiaoyu* [Cultural Changes and Bilingual Education]. Beijing: Jiaoyu kexue chubanshe.

Wright, Sue. 2004. *Language Policy and Language Planning.* New York: Palgrave.

Xie, Qihuang. 1989. *Zhongguo minzu jiaoyu shigang* [History of minority education in China]. Nanning: Guangxi jiaoyu chubanshe.

Chapter 11

Using Yugur in Local Schools: Reflections on China's Policies for Minority Language and Education

Zhanlong Ba

This is an account of two unsuccessful attempts to introduce a minority language, Western Yugur, into schools in Yugur communities in Gansu Province. As I review these two cases, my purpose is not to assess the value of teaching minority languages in public schools, but rather to analyze why these two attempts failed and what we might learn from them. I discuss in some detail the linguistic and cultural environments of Yugur communities and their impact on the trial use of Western Yugur in schools. But I also examine the interaction among national language policy, national laws, their administration, and the local communities, and conclude that lack of coherence among these four dimensions was the most important factor in the failure of the language programs. My analysis is directly relevant to a critical issue for minorities in the People's Republic of China (PRC): how to maintain minority languages in public schools where *Putonghua* (Mandarin) is often the main medium of instruction. The Yugur case may also provide insights into bilingual programs and policies in other multiethnic nations.

In this chapter, where I draw on my anthropological fieldwork, I first introduce the Yugur and their culture and review their language use. I then examine the two cases of the trial use of the Western Yugur in two schools, one in the early 1980s and a more recent one from the early 2000s. Finally, I explain how the failure of these two trials is related to the making of

China's minority language and education policies, as well as to the way these policies are implemented.

The Yugur People and their Culture

The Yugur, one of the China' ethnic groups with a small population, mainly inhabit Su'nan Yugur Autonomous County and Huangnipu Yugur Township in Jiuquan City, Gansu Province. According to the fifth national census in 2000, the total population of the people officially identified as Yugur is 13,719, ranking them forty-eighth in population size among China's fifty-five ethnic minorities.

According to traditional academic views (He and Zhong 2000; Zhong 2002, 1), the Yugur are an ethnic group with a long history. They had close relations with the *Huihe*, who in the eighth century overthrew the Turkic kingdom in the Mongolian tableland and founded the *Huihe* state (later its Chinese name was changed to *Huihu*), and with the group later identified as *Huihu*, who migrated from the northern desert to the Gansu Corridor.

The modern Yugur call themselves "Yoghur" and have a distinct culture (Qin 2005, 121–124; Yang 1996). Thus, in China's first phase of ethnic identification work during the 1950s, the Yugur were identified as a nationality for their special cultural characteristics and strong ethnic identity. They were approved by the Chinese Central Government as one of the thirty-eight initially recognized ethnic minorities. Historically, the Yugur used to believe in shamanism, Manichaeism, and Buddhism, but now they mainly believe in Gelu Buddhism from Tibet and maintain some shamanistic beliefs as well. A very few families are Christian. In Yugur communities, shamanism once meant that people performed rituals celebrating various natural phenomena (especially animals), fertility, and ancestors. When the last shaman died in the 1970s, there was no one else in the Yugur community who knew how to perform the primary religious rituals. Now the old shamanism appears only in attenuated form in various customs and traditions. Tibetan Buddhism in the Yugur communities, however, has become indigenized.

The Yugur inhabit the upland grasslands in the northern Qilian Mountains, the Gobi oasis, and the lowland meadows in the Gansu Corridor. They mainly live in three separate districts: Huangcheng Township in Su'nan Yugur Autonomous County, in the east; Kangle Township, Dahe Township, and Hongwansi Township in the center of the county; and Minghua Township in Su'nan County and Huangnipu Yugur Township in Jiuquan City in the northwest. Traditionally, the

Yugur livelihood came primarily from stockbreeding, hunting, and collecting, with farming as a supplement. For various reasons, educational development was very slow in the Yugur community, which had a high illiteracy rate before 1949. Before the late 1930s, Yugur education was generally still "life as education" and "society as school" with dependence on oral instruction. Modern Yugur schooling began after 1938 when the religious leader Gujiakanbu the seventh advised the Yugur community to initiate formal education. Small local schools and "horseback" primary schools developed slowly after 1949. Regular formal education flourished only after 1978.

Currently, the success of nine-year compulsory education in Yugur areas ranks at the top relative to other ethnic groups in Gansu Province and even nationwide among the PRC's fifty-five ethnic minorities. Specifically, according to the fifth national census in 2000, 654 in each 10,000 Yugur people have a high school education, 528 have a secondary technical school education, 362 have a junior college education, 104 have a four-year college education, and 6 are graduate students (for details, see Ba 2006; for a comparison of educational levels across the fifty-five ethnic minority groups, see Zhou 2001a). As the level of Yugur educational achievements has risen, the Yugur have come to respect teachers and education in their pursuit of material and spiritual civilization. The Yugur reached the target of "popularizing nine-year compulsory education" throughout the whole community and passed the national inspection in 1997. This significant accomplishment became one of "the ten greatest news stories about minorities in China" in 1998 (Zhong 2002, 22).

The Yugur Languages and their Use

The Yugur people use three languages, Western Yugur, Eastern Yugur, and Chinese. Belonging to the Turkic and Mongolian language groups (both in the Altai family), respectively, Western Yugur and Eastern Yugur do not have their own scripts (see Sun et al. 2007, 1759–1779 and 1925–1937). The Yugur community generally uses Chinese characters for written communication.

Because, in addition to Chinese, the Yugur use two different local languages, their linguistic environment is rather complex (CCSS 1994, 270–272). Western Yugur is used by people residing in Hongwansi, Huangcheng, Minghua, Dahe, and other towns of Su'nan Yugur Autonomous County. Eastern Yugur is used by some of the people living in Hongwansi, Huangcheng, Kangle, and neighboring areas of Su'nan

Yugur Autonomous County. Both languages are used by people in Dahe Township. Chinese is used in Huangnipu Yugur Township of Jiuquan City and also in Hongwansi, Minghua, and Dahe. During communication between the two language groups, Chinese is used relatively more often than the two Yugur languages.

Though the Yugur population has been continuously increasing, the prospect of using the Yugur languages is not promising. According to the third national census in 1982, out of a Yugur population of 10,569 in Su'nan County, 4,623 spoke Western Yugur and 2,808 Eastern Yugur (Zhong 2002, 278). However, according to the fourth national census in 1990, out of a Yugur population of 12,293 in Su'nan only 3,693 spoke Western Yugur and 3,194 spoke Eastern Yugur. From these numbers, we find that in less than a decade the number of Yugurs in Su'nan Yugur Autonomous County speaking Western Yugur was reduced by nearly 1,000 (p. 278), and the number of speakers of Eastern Yugur had risen by only a few hundred. Statistics from the Department for Ethnic and Religious Affairs of Su'nan Yugur Autonomous County show that, among the 10,079 Yugurs in the county in 1998, 5,069 spoke Western Yugur and 550 of them spoke both Eastern and Western Yugur, while 4,684 spoke Eastern Yugur, and 326 of them spoke only Chinese (Chen 2004, 13). At present, the total population of Yugur is above 16,000 at least. The persons speaking Yugur languages are only half of the total population. Furthermore, among the young, the proportion of monolingual Yugur speakers in the population is declining, while the proportions of Chinese-Yugur bilingual speakers and monolingual Chinese speakers are increasing.

There are various factors influencing the decline of Yugur-speaking population. The most important seems to be formal education (Ba 2006; Zhong 2002, 279). Yugur students study Chinese and English in schools where Chinese is the language of instruction.

At my 2004 fieldwork site, a community whose livelihood is now split between farming and herding, much of the local traditional culture of the Yugur is related to stockbreeding. Because the Yugur languages have no script, Yugur cultural traditions are mainly transmitted orally and by demonstration and imitation. Thus, oral Yugur, in this case Western Yugur, is an important element in, and the primary medium of, Yugur culture. With the decrease in the use of Western Yugur, a generation gap becomes more culturally obvious. Much of the traditional oral and intangible heritage has died out because the young generation has simply not been able, in a linguistic sense, to inherit them. Today in this community, people who can sing ancient folk songs and pass on ancient folklore using Western Yugur are already very few. When even one of such people passes away, it can be said that a library of folk culture vanishes with him or her. In my fieldwork the deepest impression that I got is that if the local people don't speak Yugur, this community has no substantive cultural difference from any farming districts of the Han majority. In addition, my investigation shows (Ba 2006) that the Yugurs in the community deeply hope that local teachers could teach Western Yugur, along with Chinese, in schools, but they are frustrated as to how to realize the goal of becoming bilingual in Chinese and Yugur and how to make their views and suggestions be understood and accepted by the local governments and schools. To remedy the situation, it seems that some mechanism for connecting the community to local schools and government is badly needed to facilitate the process of policy making and implementation of minority language education.

Two Trials of Yugur Use in Schools and Preliminary Analysis of their Failure

The Yugur communities' efforts to explore the use of Yugur languages in schools and attempts to broaden the role of native language in the local cultural repertoire are demonstrated in two educational trials of native language use. One took place in the early 1980s when China had just returned to its accommodationist minority language policies, while the other was in the early 2000s when modernization and globalization exerted extraordinary pressure on minority communities in China (for language policies changes, see Zhou 2001b; for a complete picture of bilingual education in China, see Feng 2007).

Amid a national wave of re-adoption of minority languages in schools, the first trial to teach Western Yugur as a subject took place at a school in Huangnipu Yugur Township of Jiuquan City from November 1983 to July 1984. The school hired a high school graduate from Minghai Township as the bilingual teacher. The teacher spoke Western Yugur fluently, but she had not received any training in teaching a native language as a subject or a supplementary medium in a school where Chinese was the medium of instruction. Moreover, there were no Yugur teaching materials and reference books available to her at that time. In class she covered Yugur numbers, pronouns, kinship terms, and daily expressions. She taught these materials orally, using Chinese characters as phonetic notations to write Yugur sounds.

A total of 180 students from this school participated in this trial. The students were divided into three classes: the students from grades 2 and 3 were assigned to one class, students from grades 4 and 5 to the second class, and the middle school students to the third class. The students worked very hard at learning the local language, but they had problems with their pronunciation and had to spend a great deal of class time on pronunciation drills. Nonetheless, they all made progress in their Yugur, and the youngest particularly made the most progress.

However, the Yugur classes were discontinued at the end of the school year in July 1984 and not restarted the following school year for three main reasons (Ba 1998). First, it was believed that studying Yugur negatively affected younger students' Chinese learning. Second, some parents of the Han students opposed these classes because both Yugur and Han students were required to learn Western Yugur. Third, the community lacked a supportive linguistic environment. In the community where the school is located, most Yugur people by that time had already given up speaking Yugur in favor of Chinese.

The second trial with Western Yugur took place as an extracurricular activity in Hongwan primary school from September 2003 to July 2004. On the eve of the Teachers' Day on September 8, 2003, the Party committee secretary of Su'nan Yugur Autonomous County and other county leaders met with teachers and some parent representatives about improving teaching quality. A parent representative, who was a doctor in the county hospital, raised the issue of Yugur culture and language in school with the following comments and suggestions. He pointed out that Su'nan was a multiethnic county with the Yugur as the majority, and thus it was not proper to have no Yugur culture taught in school. With the caveat that any courses on Yugur culture should not affect students' ability to progress to higher grades, the doctor suggested that the schools launch Yugur culture teaching activities. They should especially encourage Yugur students to study Yugur languages in school, most appropriately as an extracurricular activity for interested groups. He reasoned that such activities could help maintain Yugur languages, which did not have scripts and whose small speaking population was continuously declining. In addition, he said, studying Yugur languages could strengthen Yugur students' ethnic pride and motivation to learn, promoting their overall development. The suggestion was seconded by a county leader and a local cultural expert. After further discussions at the meeting, the county leaders decided to ask the county education bureau to implement a Yugur language program. Three days later, on September 11, the county education bureau disseminated a document with specific arrangements, requesting that "In order to facilitate the inheritance and development of the excellent cultures of ethnic minorities in our county and to transmit their civilizations, schools in minority communities in the county should start teaching minority languages as extracurricular activities for interested groups" (Su'nan Yugur Autonomous County 2003).

However, I learned that only Hongwan primary school actually followed the document's requirement and organized Western Yugur learning activities for a group of interested students. The group was led by a physical education teacher who could speak Western Yugur fluently but did not have any training in teaching a minority language. The teacher had access to only a few books on the native language, such as *A Short Introduction to the Western Yugur* and *Yugur Customs.* The class attracted twenty-six–forty-three students who attended voluntarily. During the group activities, Yugur kinship terms, terms for objects, and daily expressions were introduced. After a year, in September 2004, only eight upper level students remained in the group. Citing lack of sufficient student interest, the school administrators stopped their support for the group's activities.

During my fieldwork, I learned that at least five factors had a negative impact on this trial use of Western Yugur in school. First, students in the lower grades had worse performance than those in the upper grades. Second, for some of the students, there was no linguistic environment in which to maintain the target language because their home language was Chinese. Third, some students joined the group, not out of their own interest, but because their parents forced them. They did whatever their parents asked them to do, whether it was to join the group or to withdraw from it. Fourth, some language teachers opposed or did not support their students' extracurricular activities in Yugur because they thought that studying Western Yugur took students' time and energy away from studying Chinese and English. Fifth, saying that minority languages are useless and primitive, some parents did not give their children permission to join the group. They spoiled a supportive social environment for the enterprise (Ba 2005, 63–64).

Reflections on China's Minority Language and Education Policies and the Two Cases

Since 1949, China has successively promulgated, amended, and put in place several measures with implications for minority education, such as *Constitutions of the PRC*, the *Education Law of the PRC*, and *PRC's Law for Regional Autonomy for Minorities.* These laws protect in principle the rights of ethnic minorities to use and develop their own languages and writing systems, while they stipulate the promotion of *Putonghua* and standardized Chinese characters throughout the whole country (for China's minority language policies, see Zhou 2003; Zhou and Sun 2004). Policies have been made and systems have been established to ensure that these laws are carried out. However, China is a developing country, and its economy, society, culture, and education are always changing. Moreover, China has a rich social and cultural diversity. All these challenge the universalism of the PRC's laws, policies, and systems, as in the two cases I have reported here. The diversity of real life in every community brings local policy makers and implementers more difficulties and problems, which I think may be characterized in the following three ways.

First, there is what I call the "black-box effect," meaning that the public is kept in the dark about how the local government strikes a balance between locality and universality in local policy making and implementation. If all levels of local government are to be regarded as policy makers/implementers, and the public is seen as the major target of such policies, we find that the process of policy implementation from upper administrative levels to lower levels is a "black-box" to the public. All levels of local governments and their subordinate units, possessors of mainstream social discourse, and ethnic minority elites may take the initiative in the process of political "consultation" to infuse their own "will" directly or indirectly into policy. Actually, policy makers know all this, but are often frustrated because they must create a balance between the coherence and universalism of national policies and the diversity of local initiatives. Thus, policies may have more or less "flexibility" for if, when, and how to implement them, and no requirement that the decision-making process be transparent. Hence, the black-box effect.

Second, there is what I consider "stimuli-effect" in local policy making and implementation. PRC national laws and policies are not systematically and consistently enforced or implemented. Rather, local policy makers usually play the role of "firemen" and promulgate or amend and implement policies in order to handle any "new" problems, or fires that erupt in China's changing social and economic environment. The case of the trial use of Yugur at Hongwan suggests that the making and implementation of local policies for minority education may demonstrate more characteristics of a "stimulus-reaction" pattern than a coherent plan with a top-down mandate, for example. Teaching minority language and culture in schools in an ethnic autonomous area is not a new issue, but may be handled as a "new" problem because it has been newly raised by someone. The cases at hand in fact are neither unusual nor special.

Third, there is what I call "plan-effect," the idea that only what is state-planned may be implemented. Since the PRC was founded, all successive governments have emphasized the making of five-year plans for national economic and social development, and made policies to achieve the goals set in every five-year plan. It is obvious that in the tenth five-year plan (2001–2005) and eleventh five-year plan (2006–10) the Chinese government's focus has shifted to the promotion of a harmonious development for the nation's society, taken as a whole. Various social issues (such as education) in western China, where most of the ethnic minorities live, have begun to attract the attention of the government and the public. Although local governments have the authority to implement the plans according to the actual local conditions, they are more willing to accept the will of their superior governments and pursue the superior governments' plans. Thus, they overlook the stability and continuity of policies and the necessity to respond to local people's wishes, positions, and requests. For example, on the national level the maintenance of cultural diversity has been a hot topic, but locally the goals and promises of cultural development for ethnic minorities, including minority language and education, were not specified in the local autonomous government's past five-year plans, though the public appealed to preserve Yugur traditional culture. Fortunately, the central government and the State Ethnic Affairs Commission have recently requested specific programs (in new five-year plans) for minority groups with small populations, so that the Gansu Provincial Government has included, in its eleventh five-year plan, a special program for the Yugur. But the "plan-effect" remains a problem because local governments still rely on the authority of higher levels of government for specific plans.

In short, in the two unsuccessful trials we find that policy makers and implementers failed to show enough concern about the interaction among the policies, systems, and sociocultural diversity. Policies for minority language and education should not be isolated as rhetorical statements or texts, or even as "guidelines." Their implementation needs to be guaranteed by government systems and resources, support from the public, and a favorable sociocultural environment. Obviously, policies should be made and implemented in accordance with the existing laws. However, from my field observations, I find that policies sometimes and under some circumstances play a more significant role than laws. The public "habitus" is to pay more attention to top-down policies than to laws that might help transform community initiatives into long-term practice. This tendency reflects the historical role that China's government has played in initiating and enforcing social planning. To realize the goal of "governing with laws" in society at its grassroots, it is very important for local governments to act with transparency, consistency, and legality in local policy making and implementation.

Epilogue: Who is to Blame for the Failed Trials?

As a drop of water reflects the ocean, the two trials of the use of Yugur in schools are important events in the social development in the Yugur community. The two trials failed even though most people in the community value the importance of maintaining Yugur culture. Whose fault is it? Is it the fault of the schools? The policies? The government? The nation-state? Or is it simply because of modernity? Another question is whether such trials must succeed and should not be allowed to fail. All these problems deserve special attention and further research.

I argue that the approaches to maintaining and developing a minority language should be varied and implemented by all the concerned groups. The key for language maintenance and development is to keep the language alive academically, educationally, economically, and culturally through families, the community, schools, libraries, and even theme parks, the cultural and tourist industry, and so on. The two Yugur languages are priceless heritages for the Yugur people and China. No matter how difficult it is to maintain them, it is worth doing so for the sake of cultural and linguistic diversity.

From a global perspective, there is no perfect social development process in any nation-state or region. This is a basic fact of human society. The social development of China's minorities is a complicated process. So it is inevitable that the process of development encounters some errors and some setbacks; educational development is no exception to this process. We should investigate and account for the sociocultural contexts of these errors and setbacks, instead of offering only criticism. After all, before we take the next action, the more clearly we define our position, conviction, and viewpoints, the more possible it is for us to reach realistic goals and visions.

References

Ba, Zhanlong. 1998. "Xibu Yugur yu de shiyong yu jiaoxue shulue" [A Brief Account of the Use and Teaching of the Western Yugur Language], *Gansu minzu yanjiu* 18 (1): 62–64.

———. 2005. Shequ fazhan yu Yugurzu xuexiao jiaoyu de wenhua xuanze—renkou jiaoshao minzu xiangcun xuexiao jiaoyu de minzu zhi yanjiu [Community development and cultural selection for Yugur schooling: Ethnographic research on rural schooling of an ethnic group with a small population in Western China], master's thesis, Central University for Nationalities, Beijing.

———. 2006. "Yugurzu xuexiao jiaoyu gongneng de shehui renleixue fengxi" [A Social Anthropological Analysis of the Functions of Yugur], *Minzu jiaoyu yanjiu* 17 (6): 37–44.

Chen, Zongzhen. 2004. *Xibu Yuguryu yanjiu* [A Study of the Western Yugur Language]. Beijing: Zhongguo minzu shying yishu chubanshe.

Chinese Academy of Social Sciences (CCSS). 1994. *Zhongguo shaoshu minzu yuyan shiyong qingkuang* [The Status of Minority Language Use in China]. Beijing: Zhongguo zangxue chubanshe.

Feng, Anwei (ed.). 2007. *Bilingual Education in China: Practices, Policies and Concepts*. Clevedon: Multilingual Matters.

He, Weiguang, and Fuzu Zhong. 2000. *Yugurzu minsu wenhua yanjiu* [Studies of Yugur Traditional Culture]. Beijing: Minzu chubanshe.

Qin, Yongzhang. 2005. *Ganningqing diqu duominzu geju xingchengshi yanjiu* [A Study of the History of the Development of Ethnic Diversity in Gansu, Ningxia, and Qinghai]. Beijing: Minzu chubanshe.

Su'nan Yugurzu zizhi xian jiaoyuju [Educational Bureau of Su'nan Yugur Autonomous County]. 2003. *Guanyu kaizhan minzu Yuyan di'er ketang huodong de tongzhi* [Memorandum on Extra Curricular Activities for yuyan Languages]. Su'nan: Jiaofa 2003, no.199.

Sun, Hongkai, Zenyi Hu, and Xing Huang (eds.). 2007. *Zhongguo de yuyan* [Languages in China]. Beijing: Shangwu yinshuguan.

Yang, Jinzhi (ed.). 1996. *Yugurzu yanjiu lunwenji* [Collection of Studies on the Yugur]. Lanzhou: Lanzhou daxue chubanshe.

Zhong, Jinwen (ed.). 2002. *Zhongguo Yugurzu yanjiu jicheng* [A Collection of Yugur Studies in China]. Beijing: Minzu chubanshe.

Zhou, Minglang. 2001a. "The Politics of Bilingual Education and Educational Levels in Ethnic Minority Communities in China," *International Journal of Bilingual Education and Bilingualism* 4 (2): 126–150.

———. 2001b. "The Politics of Bilingual Education in the People's Republic of China Since 1949," *Bilingual Research Journal* 25 (1 & 2): 147–171.

———. 2003. *Multilingualism in China: The Politics of Writing Reforms for Minority Languages 1949–2002*. Berlin/New York: Mouton de Gruyter.

Zhou, Minglang, and Hongkai Sun (eds.). 2004. *Language Policy in the People's Republic of China: Theory and Practice since 1949*. Boston: Kluwer Academic Publishers.

Part IV

Globalizing the Discourse on Inequality and Education

Chapter 12

Affirmative Action, Civil Rights, and Racial Preferences in the U.S.: Some General Observations

Evelyn Hu-DeHart

Forty years after being launched by President Lyndon Johnson at the height of the civil rights movement in the 1960s, affirmative action in the United States is in the throes of an acrimonious debate. The upshot is that this innovative public policy is in retreat, severely battered by relentless attacks from an extreme wing of the conservative right and its many allies. Led by White male lawyers, notably Roger Clegg of the provocatively named Center for Equal Opportunity, and his many allies, including high profile African American *agent provocateur* Ward Connerly of California, these well-organized, anti-affirmative action interest groups have forced elite universities and law schools to abandon practices that overtly strive to ensure some degree of racial diversity on their campuses (discussed later in this essay) and succeeded in passing ballot initiatives in California and Michigan that ban all forms of affirmative action in the public sphere.

Ironically, the decline of affirmative action in the United States, which invented the idea, coincides with a growing and diverse number of nations around the world that looked to the United States for inspiration and thus adopted their own, unique forms of affirmative action (or "positive discrimination"). They did so to address their own urgent needs to redress historical injustice, or as a strategy to advance socially disadvantaged groups that are severely underrepresented in arenas of social mobility, notably education and employment. For example, India has proposed some form of affirmative action for the untouchables in this mass society ordered by a near-intractable Hindu caste system; Malaysia has practiced its version to help advance the *bumiputra* or "native sons"; and more closely approximating the U.S. program, affirmative action in Brazil is designed to help Brazilians of African descent compete more effectively. What they all have in common with U.S. affirmative action is the government's acknowledgment that some groups in society lag far behind a dominant group—be it high caste Hindus, white Brazilians, or Chinese Malays—in accessing and benefiting from social opportunities. To close this gap, the policy provides some kind of "preference" to those disadvantaged who demonstrate the ability and desire to compete.

If effectively implemented, the expectation is that over time, affirmative action would no longer be necessary for future generations because the historically disadvantaged would have attained the proverbial level playing field. Indeed, U.S. president Barack Obama was pointedly asked the question in May 2007 whether his two daughters should benefit from affirmative action when the time comes for them to go to college. He was the offspring of a short-lived marriage between a Kenyan graduate student (who earned a PhD in economics from Harvard) and a white mother (with her own PhD in anthropology from the University of Hawaii). Obama himself attended Columbia University and Harvard Law; his African American wife, Michelle, from a working class Chicago family whose parents did not attend college, has degrees from Princeton University and Harvard Law School. While is it impossible to document (no records are kept of affirmative action beneficiaries), given their age and race, they most likely benefited from affirmative action in obtaining admission to the highly selective Ivy League universities.[1] To the question regarding his daughters and affirmative action a generation later, Obama responded without much hesitation that his daughters "should probably be treated by any admissions officer as folks who are pretty advantaged" (Hebel 2007, A24–A25).

Even more pointed, the very fact of Obama's quick political ascendancy and subsequent viable candidacy for the presidency of the United States in 2008 has prompted some whites to ask (and some blacks Americans to fear) the question: Does the United States still need affirmative action? (Kaufman 2008, A1 and A8). In short, has Obama's quick ascent to the presidency and his own declaration that he represents the "post-race" generation of African Americans created a sharp turning point in the debate over affirmative action, to the delight, it would seem, of Clegg, Connerly et al.? But Obama has deflated whatever initial elation Connerly might have felt when he (Obama) pronounced himself in favor of maintaining affirmative action policies for the foreseeable future. He was in effect reminding fellow successful African American Connerly that if affirmative action has helped give rise to a black middle and upper class of professionals and businessmen like themselves—even if poor and working class whites suffering from job losses due to de-industrialization and globalization might well argue for some kind of government program to give them a "leg up" in life as well—legions of black and other minorities that history once abandoned are still left behind, and still demand government attention and compassion. His desire and inclination to put race aside notwithstanding, on the question of affirmative action, it seems, even Obama cannot sweep our unfinished business with race under the carpet. Here, we present a brief history of affirmation action in the United States and conclude with some suggestions to modify the policy.

The first shot heard around the United States that pierced the armor of white supremacy and launched the civil rights movement that ended legal and mandatory apartheid or racial segregation was the 1954 *Brown v. Board of Education* decision, which ruled that the prevailing system of segregated public schools was unconstitutional. It was apparent that the "separate but equal" doctrine underlying U.S. apartheid was a false promise and, indeed, a cruel hoax for black Americans, denied equal opportunities in education, employment, and other avenues of social mobility soon after liberation from slavery.[2] By the late nineteenth century, in the *Plessy v. Ferguson* Supreme Court Decision of 1896, U.S. apartheid, or legally mandated racial segregation in all facets of American life both public and private, was the law and practice of the land. Blacks and whites could not live in the same neighborhood, use the same public facilities, and perhaps most devastating of all for black prospects of social mobility, attend the same schools. In practice, this meant that African Americans, who lived in much poorer neighborhoods, owned little property, and earned much lower incomes, could never compete with white neighborhoods in establishing and sustaining good schools for their children that would prepare them for higher education, which was also segregated by law.[3]

This era in American history has conventionally been framed in terms of racial *discrimination*, highlighting, in effect, the disadvantaged in this system. Recently, however, scholars and social commentators have also examined the legacies of U.S. apartheid by focusing on those who have benefited from racial *preference*, calling attention to white Americans and the system of white privilege that has kept them on top of U.S. society since the founding of this nation. It is true that the United States was founded on the principles of liberal democracy, in which pursuit of individual freedom and the accumulation of private property constitute the basis of political and social life. In this society, the individual is paramount; merit is defined by individual achievements, so that rewards, success, and mobility are determined by the meritocracy. The first principle of a liberal democracy is the assertion that the individual and the individual's interests and well-being form the test of a good society. All this is well and good for those who are included in the system and allowed full access to all opportunities. But embedded in the founding and building of this nation is a deep and glaring contradiction between racial preference and racial exclusion. Indeed, as historian Alexander Saxton (1990) argues, the first generations of founders of the United States—Presidents Jefferson and Jackson—had set out to cultivate a "racially exclusive democracy":

> democratic in the sense that it sought to provide equal opportunity for the pursuit of happiness by its white citizens through the enslavement of African Americans, extermination of Indians, and territorial expansion at the expense of Indians and Mexicans... It is true that the United States absorbed a variety of cultural patterns among European immigrants at the same time that it was erecting a white supremacist social structure. Moderately tolerant of European ethnic diversity, the nation remained adamantly intolerant of racial diversity. (P. 10)

Under this system of white supremacy, white privilege, or a structure of legally sanctioned special advantages—racial preferences, in other words—was the norm, and no one questioned it for a long time. For one thing, those who made the laws, upheld the laws, and benefited from the laws all belonged to the same privileged group of white males. The dominant culture also reflected the white male worldview, which did not question white favoritism and, in fact, confidently declared itself a meritocracy to lend further credence to the superiority of white people who were always on top because they deserved to be there. "Let's face up to the awkward truth," historian Benjamin DeMott (1991) admonished, "special advantages are and have been for generations as American as blueberry pie" (A40).

Further solidifying white supremacy immediately preceding the civil rights era were a series of public policy initiatives aimed at a broader distribution of wealth and resources in the United States. In a new study pointedly titled *When Affirmative Action Was White* (2005), political scientist and historian Ira Katznelson presents compelling evidence that postwar progressive policies enacted by Presidents Roosevelt and Truman regarding social security, collective bargaining, and veterans' benefits for affordable housing and higher education either excluded the vast majority of African Americans or treated them differently from poor and working class whites, many of whom were lifted into the burgeoning middle class. In short, another round of racial preferences exacerbated the already large wealth and education gap between white and black. Furthermore, as more whites advanced into the growing U.S. middle class, blacks remained behind in the inner cities with their aging stock of housing and schools, while whites moved into new suburbs with gleaming new educational and recreational facilities.

It is precisely this kind of preference, favoritism, and privilege, this kind of de facto affirmative action underlying gross and persistent racial inequality, that had existed in the United States for hundreds of years until the civil rights movement in the 1960s challenged legally sanctioned and mandated racial exclusion and segregation, beginning with the public schools. The *Brown* school desegregation decision was followed a decade later by the broadly framed *Civil Rights Act* of 1964, which dismantled U.S. apartheid and outlawed racial segregation across American society. To his credit and in plain acknowledgment of America's racist past and entrenched racist legacies—and perhaps from personal experience as a privileged white southern male—President Johnson knew that more action beyond changing the laws was necessary to address the uneven playing field of white privilege and black exclusion. The following year, he issued *Executive Order 11246* ordering public institutions to take "affirmative action" to ensure that members of groups excluded from equal opportunity in the past—meaning in his mind primarily African Americans—were given every chance to complete fairly. Johnson's honest insight would be confirmed by other white males, including those on the Kerner Commission of 1968, which studied the causes of the devastating race riots of Watts (California) and concluded that, if left to their own devices, white Americans would follow "business as usual" and discriminate against blacks and other minorities as they had been doing for hundreds of years (Wilkins 1995). To make affirmative action more politically palatable, Johnson extended the policy to cover other officially recognized minorities and all women, in so doing acknowledging and addressing the legacies of gender discrimination that had also pervaded U.S. society from its inception.[4]

In theory and practice, affirmative action requires agencies and institutions to make every effort to seek out qualified candidates from every possible source—but to focus especially on individuals from previously excluded and, hence, currently underrepresented groups—to compete for limited resources in education, employment, and business (government contracting and small business loans). The point of affirmative action was simple: to deliberately attack and break down barriers that had in the past systematically barred blacks and other minorities (and in some cases women, as well) from equal opportunity. In this sense, affirmative action was designed to be a top-down strategy meant to level the playing field for those deemed capable of taking advantage of opportunities no longer denied them because of their race or gender. Contrary to popular myths, it is not about racial quotas, or about giving jobs, educational opportunities, or government contracts to unqualified individuals incapable to making good use of them for upward mobility and self-advancement. Affirmative action was not conceived as an antipoverty strategy, hence not aimed at economic inequality as such; of course, when properly applied, beneficiaries of better educational and employment opportunities should experience significant upward social mobility. And in as much as race and class do intersect in a profound and serious way, issues of class have ineluctably become a part of affirmative action discussions, but race and class are not interchangeable categories and hence cannot substitute for each other. Putting it bluntly, affirmative action does not obviate the need for antipoverty programs, and racial discrimination adversely affected rich and poor minorities.

Designed to address the legacies of U.S. racism, affirmative action is based on *race-conscious* strategies.[5] As such, these correctives are necessarily group-based because, as long time U.S. Civil Rights commissioner and historian Mary Frances Berry reminds us, "discrimination is group-based" (quoted in Minzesheimer 1995, A4). It can also be argued that affirmative action can help smooth the transition from a blatantly racist past to a potentially color-blind future, but only if taking account of race constitutes part of the transitional strategy. In the trenchant words of Justice Harry Blackmun of the U.S. Supreme Court: "In order to get beyond racism, we must first take into account race, and in order to treat some persons equally, we must treat them differently" (quoted in Editorial, *New York Times* 1996, A14). Or as civil rights leader Martin Luther King put it more bluntly, "A society that has done something special against the Negro for hundreds of years must now do something special for the Negro" (quoted in Rockwell 1996, 13). To put it most directly, affirmative action works only when it takes account of the individual's race, thereby operating in a race-conscious way, to ensure that qualified minorities are well-informed of opportunities they are eligible for, and have equal access to them. The emphasis is on leveling the playing field for individuals to compete for limited resources. By forcefully addressing the legacies of past discrimination, affirmative action for many proponents rests on the ideal of restitutive or restorative justice. In practice, it becomes a form of racial preference.

In this regard, affirmative action goes beyond antidiscrimination laws and therefore is not a simple duplication of, nor can be replaced by, such laws. In reality, the civil rights of Americans, even after passage of the *Civil Rights Act* in 1964, are routinely flouted and ignored. Minority individuals experiencing discrimination rarely press charges against wealthier, more powerful, better connected superiors or employers, for the simple reason that such actions are too complex and costly for most of them to undertake.

Only the most egregious and well-publicized violations receive donated or public resources to seek redress; most incidences of discrimination are too mundane to attract attention, thus go unattended, the victims left to nurse their own wounds.

Over time, because affirmative action has opened up opportunities for many more Americans at a time of broad consensus that the time has come to end discrimination and embrace diversity, greater competition for scarce resources has inevitably invited intense scrutiny of the program and evoked difficult questions, such as: How should society distribute limited resources, such as higher education? By what means does society ascertain how those who receive such benefits make best use of them? And for what greater social good? Thus for many, the crux of affirmative action has also come to rest on the ideals of distributive justice, looking to the future as much as the past. If training a black doctor may mean a new clinic in the underserved community of largely black and poor East Los Angeles, or training a Native American doctor committed to working on her even more underserved reservation, does that justify taking race into consideration in the medical school's admissions process when there are many more qualified applicants than available space can accommodate? These were precisely the difficult issues that the U.S. justice system wrestled with early in the history of affirmative action when it became clear that white males, who had historically faced practically no competition from qualified minorities, no longer enjoyed exclusive claim to all desirable social resources. In the 1978 *Bakke* case, when the University of California at Davis Medical School faced just such a challenge from a white male applicant initially denied admissions, the Supreme Court narrowed the parameters of affirmative action to disallow institutions from setting aside a certain percentage of seats for underrepresented minorities, but upheld the use of racial preferences in appropriate circumstances and under strict scrutiny.[6]

The question still remains: After forty years, exactly what has affirmative action achieved for black and other minorities in America, and for women in general? How much and for whom? From various studies and statistics gathered, some clear answers have emerged. Although affirmative action was an immediate and direct outgrowth of the Civil Rights Act of 1964 that outlawed legal discrimination against black Americans, then the largest minority group at about 13 percent, it was quickly expanded to other racialized groups, that is, Mexican Americans and Puerto Ricans, American Indians, Chinese Americans, Japanese Americans, Filipino Americans.[7] Furthermore, in 1968, as a result of intense lobbying by feminist and women's rights groups, and some say in order to make affirmative action politically more palatable by soft-pedaling the centrality of race, affirmative action coverage was extended to all women, as well, in recognition of historical gender discrimination. Consequently, well over 50 percent of U.S. society—close to 70 percent by the twenty-first century—is covered by affirmative action. This fact alone makes it at least understandable why the only group left out, white males, have become increasingly uncomfortable about the project.

All studies conclude, and statistics easily confirm, that white women have benefited the most from affirmative action. By the end of the twentieth century, white women came to constitute 45 percent of the workforce, compared to 7.5 percent for Latinos, 10 percent for African Americans, and 2.6 percent for Asian Americans. Although still far from attaining parity with white males, and still facing serious glass ceiling barriers, white women have risen more rapidly and spread more evenly across the workforce than any racial minority group. For example, in administrative jobs between 1960 and 1990, white women rose from 9 to 33 percent of the total; in professional positions, from 22 to 34 percent; and in technical fields, from 14 to 30 percent (Connell and Nazario 1995, A1, A34–A36). In Congress and many state legislatures, white women are no longer a novel sight, and perhaps most impressively, white women are now presidents at elite higher education institutions, including several Ivy League universities (Princeton, Pennsylvania, Brown), the Massachusetts Institute of Technology (MIT), major state research universities (Michigan, Ohio), as well as the Ford Foundation.[8]

This should come as no surprise, for white women have some clear advantages over minorities given the fundamentally moderate design of affirmative action. There are more of them; as a group they are better educated and suffer no major educational deficits or disadvantages through high school and college[9]; collectively, they are less mired in poverty and well-represented in the middle and upper classes. Culturally and racially, they are more accepted by white males, who hire, sire, and marry them. As Professor Ruth Rosen (1996) acknowledges about professional women like herself: "White middle class women were best positioned to take advantage of affirmative action programs. Once the barriers were lifted, we leaped into male-dominated professions and occupations" (B5). Rosen points out the fundamentally moderate approach of affirmative action, for it can only make available opportunities to already qualified individuals, whether women or minorities. Given their social, class, economic, and educational backgrounds that more closely approximate the criteria of meritocracy set up by white males, and that affirmative action accepts without question as fair and universally valid, white women in general more readily meet these standards of qualification. White women's progress has outpaced all minority men and women in every sector: the workforce, management, and higher education (as students, faculty, and administrators).

When affirmative action in higher education was seriously challenged for the second time at the end of the twentieth century, white women were chosen as plaintiffs alleging wrongful denial of admissions because "less qualified" minorities were admitted instead. They even coined a phrase borrowed from the civil rights movement to describe their grievance: reverse discrimination. By design, white women were recruited to be lead plaintiffs in the two major cases challenging the use of racial preferences in higher education: *Hopwood v. Texas* (1992) and *Grutter v. Bollinger* (1996) (see Spann 2000). Denied admission to the University of Texas Law School and the undergraduate college of the University of Michigan, respectively, Cheryl Hopwood and Barbara Grutter alleged that their rightful places had been given to less qualified minority male applicants. It would seem that for some white women, affirmative action has been recalibrated not so much along strictly male-female gender lines as along white female-minority male lines. So while white women have been competing with minority groups, especially minority men, no one is seriously challenging white men, the group that has retained firm control of the best, most powerful, and most lucrative positions in industry, education, and government. As columnist DeWayne Wickham (1995) noted in the heat of these attacks against affirmative action and racial preferences, little in the end had changed about the command structure of U.S. society, for white males held 95 percent of industry's top jobs, 80 percent of tenured faculty positions, and 90 percent of U.S. Senate seats (A11).

The strain between white women and minority men is both real and perceived, leading to pointed questions such as: Are white women squeezing out minority men? Another questions also begs to be asked: Why has affirmative action been almost exclusively framed around "racial preference," with hardly a whisper about "gender preference," when clearly women have benefited disproportionately? Is white America ready to discontinue affirmative action now that it has taken care of its own women and granted them more privileges and opportunities? Why is white America more receptive, more comfortable, about affirmative action for white women than for minority men and women, as a 1995 CNN/Gallup poll uncovered? (Johnson and Moss 1995, A6). Has affirmative action, which started out as a modest measure proposed by a southern president at the height of the civil rights era to help address racial inequalities produced by centuries of white supremacy, become another way for Americans to talk about *race*, the unfinished business of the United States that black intellectual W. E. B. Du Bois predicted at the dawn of the twentieth century and that Swedish sociologist Gunnar Myrdal confirmed in mid-century?[10]

Further clouding the affirmative action picture is the disproportionate progress made by one minority group in the past two decades. When Asian

Americans were first included in affirmative action, their numbers were extremely small, and they were truly invisible in all institutions and sectors of U.S. life. Since the 1970s, with renewed and massive immigration from all parts of Asia, their numbers have increased dramatically, at ten million strong in the early twenty-first century, a tenfold increase from the mid-1960s. Led mainly by Chinese, Japanese, Koreans, and South Asians, Asian Americans have made impressive gains in American higher education, becoming an "overrepresented" minority because their numbers at elite institutions far outnumber their 5 percent proportion of the population. Asian achievements cannot be wholly explained by cultural factors such as discipline, hard work, and commitment to formal education, for many of these highly motivated immigrants come also with considerable social, human, and economic capital. Higher educational attainment means that Asian Americans usually out-compete other minorities and even whites in high tech and white collar professional jobs, although they remain grossly underrepresented at the top leadership echelons of industry and education, and almost totally absent in politics. One of their responses to the glass ceiling is to leave a structured work environment to form their own businesses, be their own bosses (Hu-DeHart 2008).

In light of the markedly differential affirmative action experiences of Asian Americans and white women on one side, and blacks, Latinos, and Native Americans who remain severely underrepresented in education on the other, is it not time to rethink the meaning of affirmative action for each of their original intended beneficiaries, and to re-conceptualize the future of affirmative action overall?

To begin with, the groups need to be disaggregated, and the strategies particularized. For white women and most Asian Americans (there are some exceptions among the predominately Southeast Asian refugee communities), the challenge ahead appears no longer primarily one of access to opportunities, but rather of how to reach for and break through the glass ceiling. Strategies should focus on upward mobility and advancement within institutions once entry has been gained.

For African Americans, Latinos, and Native Americans, the challenge today and into the foreseeable future remains what has always been—that of meaningful and significant access to and representation in education and other major institutions. For them, traditional affirmative action based on proactive racial preference remains urgently needed.

We end this brief discussion with an observation made at the beginning: that countries around the world have devised their own models to address their own conditions of historical inequities. Some of these have been inspired by U.S. affirmative action, but are usually based on some form of group approach, as opposed to the legal rights of the individual. For the interests of this volume of essays on the theme of minorities and education in China today, one might well ask if the time has come for China to revise its version of affirmative action, particularly in light of Minglang Zhou's recent observation that China has adopted a new approach to citizenship for non-Han minorities, which he describes as "one nation with diversity." He argues that in this approach, "the state gives non-Han minorities full access to citizenship," and expects them to assume all the duties of political, social, and cultural citizenship (Zhou 2008, 8). That given, since these non-Han minorities lag far behind the Han majority in accessing and benefiting from political, social, and cultural opportunities, might not some kind of individual-based affirmative action modeled after that in the United States be in order in China as well?

Notes

1. When asked directly if he received affirmative action consideration in admission to Harvard Law School and election to chief editor of the *Harvard Law Review*, Obama did not dodge but answered forthrightly: "I have no way of knowing if I was a beneficiary of affirmative action.... If I was, then I am certainly not ashamed of the fact, for I would argue that affirmative action is important precisely because those who benefit typically rise to the challenge when given an opportunity" (Kaufman 2008, A1 and A8).
2. In the West and Southwest, Asian Americans and Mexican Americans were also forced into segregated public schools, while Native Americans experienced their own form of segregated schools on the reservations and in government boarding schools. For Asian Americans, see Evelyn Hu-DeHart (2004) and for Mexican Americans, see Marco Portales (2004). For American Indians, see David W. Adams (1995).
3. Racial segregation in the United States was so comprehensive that it severely regulated love and romance and outlawed interracial marriages; for a recent discussion on the politics of interracial intimacy in the decade just preceding the *Brown* decision, see Lubin (2005).
4. Of course, gender discrimination and racial discrimination were different systems of oppression, although they share some fundamental characteristics, the key one being the group-based nature of such forms of discrimination. Historically, however, there has always been a strong correlation and intersection between race and class for racial minorities, while women as a category cut across both race and class. Specifically, while white women suffered from gender discrimination, they had access to race and often class privileges, which could mitigate the negative effects of gender.

5. For the sake of a smoother discussion, and because this volume is about minority rights, we will limit the bulk of our discussion on affirmative action mostly to race, while keeping in mind that, as already noted, the policy was extended to cover all women. But later in this essay, when we assess the achievements of affirmative action, we will return briefly to the question of affirmative action coverage for women.
6. Much has been written about the landmark Bakke case. A good discussion can be found in Girardeau A. Spann (2000, 15–18).
7. When affirmative action was expanded to cover other minority groups, Latinos consisted primarily of Mexican Americans and Puerto Ricans, and Asian Americans consisted mainly of Chinese, Japanese, and Filipino Americans, but the composition of these two groups had begun to undergo dramatic demographic changes in terms of both numbers and internal diversity, due to the 1965 immigration reform that once again opened the doors of the United States to Asian and Latin American/Caribbean immigrants after decades of exclusion.
8. With the exception of Ruth Simmons of Brown University—a descendant of slaves—all the other women presidents at these elite institutions are white.
9. Indeed, by the twenty-first century, more women than men graduate with a college degree in the United States; this is true in the white community, but the gender imbalance has become critically acute in the black and Latino communities, a matter of growing concern to educational and political leaders. In this regard, one might speculate that somehow, affirmative action might have played a role in promoting women of color in education, even if the causes of low male minority participation in higher education are multiple and complex, the solutions well beyond the limited scope of affirmative action.
10. W. E. B. DuBois declared that "the problem of the twentieth century is the problem of the color line," in his classic book *The Souls of Black Folk*, first published in 1903 and having since enjoyed many editions, the latest by Bedford Press in Boston in 1997; Myrdal (1944).

References

Adams, David W. 1995. *Education for Extinction: American Indian Boarding School Experiences, 1875–1928*. Lawrence: University of Kansas Press.

Connell, Richard, and Sonia Nazario. 1995. "Affirmative Action. Fairness or Favoritism? How Well Does It work?" *Los Angeles Times,* September 10: A1, A34–A36.

DeMott, Benjamin. 1991. "Legally Sanctioned Special Advantages Are A Way of Life in the United States," *The Chronicle of Higher Education*, February 27: A40.

Dubois, W. E. B. 1903. *The Souls of Black Folk*. Chicago: A.C. McClurg.

Editorial. 1996. "Bad Law on Affirmative Action," *New York Times,* March 22: A14.

Hebel, Sara. 2007. "An Interview with Barack Obama: 'The Most Important Skill is Knowledge,'" *The Chronicle of Higher Education,* November 16: A24–A25.

Hu-DeHart, Evelyn. 2004. "An Asian American Perspective on *Brown,*" in *The Unfinished Agenda of Brown v. Board of Education*, eds. Black Issues in Higher Education, 108–122. Hoboken: John Wiley.

———. 2008. "Asian Americans and Academic Achievement," unpublished paper available upon request from author.

Johnson, Kevin, and Desda Moss. 1995. "Affirmative Action Debate Skips Women," *USA Today,* February 28: A6.

Katznelson, Ira. 2005. *When Affirmative Action was White. An Untold Story of Racial Inequality in Twentieth Century America.* New York: Norton.

Kaufman, Jonathan. 2008. "Fair Enough? Barack Obama's Rise Has Americans Debating Whether Affirmative Action Has Run its Course," *The Wall Street Journal,* June 14–15: A1 and A8.

Lubin, Alex. 2005. *Romance and Rights: The Politics of Interracial Intimacy, 1945–1954.* Jackson: University Press of Mississippi.

Minzesheimer, Bob. 1995. "Affirmative Action Under Fire," *USA Today,* February 23: 4A.

Myrdal, Gunnar. 1944. *An American Dilemma: The Negro Problem and American Democracy.* New York: Harper.

Portales, Marco. 2004. "A History of Latino Segregation Lawsuits," in *The Unfinished Agenda of Brown v. Board of Education*, eds. Black Issues in Higher Education, 124–136. Hoboken: John Wiley.

Rockwell, Paul. 1996. "The GOP Misquotes Martin Luther King," *San Francisco Examine*, March 18: A13.

Rosen, Ruth. 1996. "More Than Ever, UC Needs Goodwill," *Los Angeles Times,* July 24: B5.

Saxton, Alexander. 1990. *The Rise and Fall of the White Republic.* London: Verso.

Spann, Girardeau. 2000. *The Law of Affirmative Action: Twenty-Five Years of Supreme Court Decisions on Race and Remedies.* New York: New York University Press.

Wickham, DeWayne. 1995. "Clinton Teeters on High Wire," *USA Today,* March 6: A11.

Wilkins, Wilkins. 1995. "Racism Has its Privileges: The Case for Affirmative Action," *The Nation,* March 27: 409–416.

Zhou, Minglang. 2008. "Models of (Multi)Nation State Building and the Meaning of Being Chinese in Contemporary China," paper presented at the Critical Han Studies Conference & Workshop, Stanford University, Palo Alto, CA, April 25–27, 2008.

Chapter 13

Learning about Equality: Affirmative Action, University Admissions, and the Law of the United States

Douglas E. Edlin

Affirmative action is one of the most divisive and contentious legal and policy issues in the United States. It would have been difficult to predict the path this issue would take, however, given its somewhat inconspicuous beginnings. Affirmative action entered U.S. politics legally and linguistically with the signing of Executive Order 10,925 by President John F. Kennedy.[1] This Order was intended to prohibit racial discrimination in the hiring and contracting practices of the federal government. The pertinent language reads: "The contractor will not discriminate against any employee or applicant for employment because of race, creed, color, or national origin. The contractor will take *affirmative action* to ensure that applicants are employed, and that employees are treated during employment, without regard to their race, creed, color, or national origin" (Kennedy 1961; emphasis added). In the decades that followed, affirmative action would come to be seen by many in the United States as tangible evidence that the government is committed to achieving substantive social justice and racial equality. At the same time, affirmative action is also viewed by many others in the United States as fundamentally inconsistent with the notion that the U.S. Constitution is, in Justice Harlan's famous phrase, "color-blind" (*Plessy* 1896, 559).

As the pitch of the arguments on both sides has become more strident, the public debate in the United States about affirmative action continues to generate much more heat than light. In this chapter, I will outline the legal development of affirmative action in the Supreme Court of the United States and survey the strongest legal and policy arguments on both sides of the issue. My hope is that this will, at least, serve as a useful introduction to the issue for those who may be somewhat unfamiliar with U.S. law and politics. My hope is also that those who are all too familiar with the issue in U.S. law and politics might use this chapter as an opportunity to review the arguments on each side. In addition, I offer some comparisons and contrasts, primarily in the notes, between the experiences of China and the United States in relation to the conception and implementation of positive action policies in the two nations. In China and the United States, a unifying feature of the legal and policy debate is that these policies are understood as efforts to achieve equality, although in the United States that is often conceived in individuated terms while in China the focus seems geared more toward group integration and national identity.[2]

Before I begin, I should mention a few familiar arguments in the affirmative action debate that will not be seen here. First, I do not engage in any discussion of meritocracy. The contested questions of what constitutes merit and how it might be measured run equally through both sides of the debate. Proponents of affirmative action argue that the notion of merit is itself racially defined and manipulated (Bowen and Bok 1998, 276–278). Opponents of affirmative action respond that attempts to relativize the concept of merit are merely meant to camouflage the fact that minority students often present lower test scores and grade point averages than other demographic categories of student (Thernstrom and Thernstrom 2002, 185–186). I will discuss the issue of standardized test scores briefly later, but I will not review the more general discussions of merit in this chapter. Second, I do not discuss the argument, advanced by Justices Scalia and Thomas, that affirmative action is a social policy fundamentally inconsistent with the original understandings of the framers of the Fourteenth Amendment (*Adarand* 1995, 240; *City of Richmond* 1989, 528). Properly understood, this version of originalist opposition to affirmative action is demonstrably incorrect as a matter of historical fact (Rubenfeld 1997, 430–432; Schnapper 1985, 755–788; Sunstein 1999, 127). As a result, I avoid this argument because it will only confuse an already complicated legal landscape. Third, I do not address the argument that affirmative action attracts and admits minority students to highly competitive institutions where they are unprepared for the necessary level of academic rigor and are inevitably unable (or unlikely) to succeed. Although frequently made, this claim is empirically inaccurate (Kane 1998, 18–19, 22–23; Lempert et al. 2000).

Finally, in the United States, affirmative action arises in the context of admissions to schools of higher education and in connection with practices of hiring, retention, and promotion in various employment settings. For purposes of this chapter, I will concentrate almost exclusively on affirmative action in admissions decisions rather than in employment practices. Although it is necessary for me to cite to judicial decisions in both categories, the textual discussion is limited to the educational setting.

Affirmative Action in the Supreme Court

One of Alexis de Tocqueville's (1945) most famous observations of the U.S. governmental system was that "scarcely any political question arises in the United States which is not resolved, sooner or later, into a judicial question" (280). Whatever the merits of Tocqueville's view of U.S. legalism (or litigiousness) more generally, his words could not more accurately describe the treatment of affirmative action as a political question. Almost immediately after it entered fully into the political consciousness of the nation, it was challenged in court as a violation of the Equal Protection Clause of the Fourteenth Amendment.[3]

Bakke

The first Supreme Court case in which affirmative action in higher education was addressed is *Regents of the University of California v. Bakke.* Allan Bakke, a white man, applied twice (in 1973 and 1974) for admission to the University of California at Davis Medical School. His application was denied. Mr. Bakke objected to his rejection on the grounds that, in both years in which his application was rejected, other minority applicants were admitted with lower grade point averages, MCAT scores, and benchmark scores in the UC Davis application review system. After his second rejection from the UC Davis Medical School, Bakke filed a lawsuit in California state court claiming that the Medical School's admissions program discriminated against him on the basis of his race and challenging the constitutionality of the Medical School's admissions procedure. Bakke's specific challenge related to the Medical School's "special admissions program," which created a separate admissions track and a separate admissions committee for members of minority groups (specifically referencing "Blacks," "Chicanos," "Asians," and "American Indians"). Bakke alleged that the special admissions program violated the Equal Protection Clause of the Fourteenth Amendment to the U.S. Constitution, Title VI of the Civil Rights Act of 1964 and the California Constitution (*Bakke* 1978, 274–278).

The California trial court denied Bakke's claim, but the trial court's decision was reversed on appeal to the Supreme Court of California and the California Supreme Court's decision was appealed to the Supreme Court of the United States. Justice Lewis Powell wrote the Court's main opinion. The University of California offered four fundamental goals of its affirmative action plan: (1) to increase minority presence and reduce the historic deficit of traditionally disfavored minorities in medical schools and the medical profession, (2) to remedy past discrimination and counter the effects of long-standing societal disadvantage visited upon certain minority groups, (3) to enrich minority communities by increasing the number of physicians who will practice medicine in generally underserved areas, (4) to increase diversity in student populations and realize the educational benefits that flow from the exchange of varying student backgrounds, experiences, perspectives, and opinions (305–306). Justice Powell addressed each of these goals in turn.

In response to the stated goal of increasing minority presence, Justice Powell concluded that this goal was facially invalid, because the U.S. Constitution prohibits favoring one group of citizens over another solely on the basis of their race or ethnicity (307).

In terms of the effort to remedy past discrimination, Justice Powell indicated that this might be an acceptable basis for an affirmative action program, but not in the *Bakke* case, because the University had not and could not pursue the necessary inquiry into the existence or effects of systematic discrimination on minority medical school applicants. Indeed, Justice Powell suggests that no educational institution could ever adequately pursue this task and effectively eliminated remedying past discrimination as a potential justification for affirmative action programs in the educational setting (309–310).[4]

Where enriching underserved minority communities is concerned, Justice Powell doubted that the program could achieve this goal. No matter what the incoming medical students might state or believe about where they will ultimately go on to practice medicine, there is no guarantee that these communities will ultimately receive the benefits of these doctors' education and expertise and no way for the University to compel the doctors' compliance with their stated intentions upon graduation from medical school. In any event, in the *Bakke* case, there was no evidence in the record that UC Davis had or intended to pursue any measures to encourage minority physicians to practice in minority communities. As a result, Justice Powell discounted this potential goal of the affirmative action program (310–311).

Increasing diversity in the student population was the sole justification for an affirmative action program in the education setting that Justice

Powell upheld as constitutional (311–312). A constitutional basis for the diversity justification is found, according to Justice Powell, in the principle of academic freedom traditionally protected by the First Amendment. In pursuance of this principle, institutions of higher education may attempt to provide and foster an educational atmosphere of "speculation, experiment and creation" through "a robust exchange of ideas" (312). Moreover, the principle of academic freedom encompasses the independent judgment of colleges and universities that efforts to increase diversity through affirmative action assist in creating this atmosphere and contribute to the exchange of ideas central to their overarching educational mission (312–314).

Two other aspects of Justice Powell's opinion are central to the ensuing legal and policy debate about affirmative action in the United States. First, a crucial aspect of the UC Davis admissions plan for Justice Powell was its use of a quota system. Sixteen of the one hundred seats in the incoming UC Davis Medical School class were reserved for minority applicants and white applicants were precluded from competing for these seats (274, 288–289). Four justices believed that the UC Davis quota program was constitutional (373–374, 378–379). Justice Powell concluded, however, that this undifferentiated consideration of race in admissions was unsupportable under the Constitution and under the promotion-of-diversity rationale for affirmative action, because the quota system promotes diversity in abstract and absolute terms, rather than as a means to achieve the legitimate educational goal of improving debate and understanding (315). However, while the specific affirmative action quota system implemented by UC Davis in *Bakke* was held unconstitutional, Justice Powell went on to explain that other affirmative action programs, such as Harvard University's "plus-factor" plan, are constitutionally acceptable insofar as they treat race as a factor relevant to the evaluation of an individual student's profile rather than as the factor that fully determines a particular student's admissions decision (316–317, 321–324).

Second, members of the *Bakke* Court were divided over the appropriate standard of review to apply in cases of "benign" racial classifications. Four justices believed that, as a system of racial classification intended to benefit rather than to disadvantage the relevant minority group, the intermediate scrutiny standard should apply to affirmative action programs (358–359).[5]

In the end, Justice Powell occupied (by himself) the middle ground in his *Bakke* opinion. He disagreed with the four dissenting justices' argument that the Constitution prohibited any consideration of race whatsoever in admissions decisions (297). He also disagreed with the four concurring justices that intermediate rather than strict scrutiny should apply to affirmative action programs and that a quota system such as the UC Davis plan was constitutional (290, 295–297). Instead, Justice Powell decided that a race-based affirmative action program could be upheld under the strict scrutiny test, provided that the program evaluated each applicant individually and did not provide any absolute advantage in the admissions process solely on the basis of race.

Grutter

Twenty-five years after deciding *Bakke*, the Supreme Court revisited the constitutionality of race-based affirmative action programs in higher education. In the years after *Bakke* was decided, two lower federal courts reached conflicting decisions about the ongoing validity of Justice Powell's decision and, therefore, about the ongoing vitality of race-based affirmative action programs in university admissions (*Hopwood* 1996; *Smith* 2000).

In *Grutter v. Bollinger*, the Supreme Court noted the conflict among certain federal circuit courts concerning the *Bakke* decision and settled (for the time being) the question whether *Bakke* remained the law and whether colleges and universities could continue to consider race as a factor in their admissions decisions. In doing so, the Court reviewed the affirmative action plan used by the University of Michigan Law School, which was based explicitly on the Harvard plus-factor plan upheld in *Bakke* (*Grutter* 2003, 321, 337).[6] Barbara Grutter is a white Michigan resident whose application for admission to the University of Michigan Law School was denied. Ms. Grutter's core complaint against the Law School was that "the Law School uses race as a 'predominant' factor, giving applicants who belong to certain minority groups 'a significantly greater chance of admission than students with similar credentials from disfavored racial groups' " (317).

Writing for the majority, Justice O'Connor denied Ms. Grutter's claims and upheld the Law School's use of race as a plus-factor in its admissions decisions. Justice O'Connor began her analysis by reaffirming Justice Powell's conclusion in *Bakke* that strict scrutiny is the standard of review applied to all classifications based on race, benign or otherwise. According to this standard, any classification based on race must be "narrowly tailored to further compelling governmental interests" (326). Where affirmative action in education is concerned, Justice Powell's opinion in *Bakke* indicates that the only public interest sufficiently compelling to justify distinguishing applicants on the basis of race is the achievement of diversity among the student population (324; *Bakke* 1978, 311).

The next question, then, was whether the use of a race-based affirmative action program by the Law School was narrowly tailored to achieve its compelling interest in diversity. Again, the answer given by the *Grutter* Court was yes. The basis for this ruling was that the Harvard plan employed by the Law School necessitated "truly individualized consideration" that did not "insulate applicants who belong to certain racial or ethnic groups from the competition for admission" (*Grutter* 2003, 334). The *Grutter* ruling prohibits colleges and universities from establishing set numbers of seats for applicants from defined minority groups and it prevents the establishment of separate admissions processes or tracks for certain categories of candidate. The *Grutter* ruling permits the use of race as a plus-factor in the admissions calculus, however, so long as race is considered in the same manner as many other plus-factors (e.g., geographic diversity, legacy status, athletic achievement, artistic ability, etc.) in creating an individualized profile of each applicant and maintaining balanced competition among all applicants for all class seats (334).

One of the key aspects of the *Grutter* opinion is its consideration of the argument that race-based affirmative action plans, even more flexible plans such as the one used by the Law School, cannot satisfy strict scrutiny because there are always race-neutral means of achieving the goal of diversity in the student body. If race-neutral methods are available, then by definition a race-based plan is not narrowly tailored enough to satisfy constitutional standards. After noting that the Law School seriously considered several race-neutral alternatives, the Court explained that these race-neutral alternatives (such as a lottery system or decreasing emphasis on GPA and LSAT score for all applicants) would effectively force the Law School "to choose between maintaining a reputation for excellence or fulfilling a commitment to provide educational opportunities to members of all racial groups" (339). According to the Court, neither the Constitution nor the strict scrutiny standard requires a university to make this choice. Instead, the Law School is entitled to maintain its educational interest in increasing the racial diversity of its student population along with the commitment to academic rigor and selectivity that are central to its educational mission and its reputation for excellence (339–340). Unlike the race-neutral alternatives, the Harvard plan allows the Law School to balance and achieve these goals in the manner it has determined is best for the institution and its students.

In *Grutter*, as in *Bakke*, the Supreme Court ruled that institutions of higher education have a compelling interest in achieving diverse student populations and held that certain race-based affirmative action admissions programs are narrowly tailored to achieve this interest. In reaching this conclusion, the Court addressed the Law School's stated goal of achieving a "critical mass" of minority students within its overall population. This element of the case was crucial, because Ms. Grutter argued that critical mass was simply a euphemism for the sort of quota set-aside system that was invalidated in *Bakke.* Justice O'Connor concluded, as had Justice Powell before her, that the Harvard plus-factor plan used by the Law School in *Grutter* avoided the concerns raised by the UC Davis quota system at issue in *Bakke* for one dispositive reason:

> Here, the Law School engages in a highly individualized, holistic review of each applicant's file, giving serious consideration to all the ways an applicant might contribute to a diverse educational environment. The Law School affords this individualized consideration to applicants of all races. There is no policy, either *de jure* or *de facto,* of automatic acceptance or rejection based on any single "soft" variable... [T]he Law School awards no mechanical, predetermined diversity "bonuses" based on race or ethnicity. Like the Harvard plan, the Law School's admissions policy "is flexible enough to consider all pertinent elements of diversity in light of the particular qualifications of each applicant, and to place them on the same footing for consideration, although not necessarily according them the same weight." (337)[7]

According to *Grutter,* so long as each applicant is evaluated as an individual in the admissions process, the racial identity of that individual is one factor, among many, that may be considered when making an admissions decision.

Grutter and *Bakke* are the two leading cases that established the law of the United States regarding affirmative action. Now that we have reviewed the legal rulings in these two cases, we can turn to the different policy and legal arguments made on both sides of the issue.

The Affirmative Action Debate

Arguments against Affirmative Action: Reinforcing Stereotypes

A frequent and powerful argument against affirmative action is that it simply reinforces the stereotype that certain racial and ethnic minorities cannot succeed without some special assistance beyond individual merit and accomplishment (*Bakke* 1978, 298). As Stephen Carter (1991) puts it, "the durable and demeaning stereotype of black people as unable to compete with white ones is reinforced by advocates of certain forms of affirmative action" (50). The argument here is that affirmative action actually reflects ingrained notions of white superiority and minority inferiority (52–62). For this reason, affirmative action programs are sometimes said to stigmatize minority students during and after their educations (Lawrence and Matsuda 1997, 126–129).

As a result of affirmative action programs in admissions, those who do not benefit from the programs assume that minority students were accepted "just because they are minorities." The underlying assumption is that the beneficiaries of affirmative action programs could not have gained acceptance in the absence of the program. Accordingly, rather than mitigate the effects of ingrained and institutionalized stereotypes by increasing diversity, encouraging dialogue, and broadening understanding between different ethnic and racial groups, affirmative action actually further entrenches the assumptions that ingrained and institutionalized the stereotypes in the first place (Krieger 1998, 1263). This results in what has been called "attributional ambiguity" (1266). Simply put, this means that the existence of affirmative action always permits certain people to doubt that the innate abilities or accomplishments of minority candidates were the principal basis for their admission to an institution of higher learning.

Arguments against Affirmative Action: Harming Innocents

This point was a particular concern of Justice O'Connor in *Grutter*. Although she determined that, at the present time, race-conscious admissions policies do not necessarily cause undue harm to innocent nonminority applicants, Justice O'Connor also noted that, in her opinion, undue harm would be caused by these programs if they continued indefinitely. Consequently, Justice O'Connor built a twenty-five-year sunset provision into her *Grutter* opinion in an effort to prevent undue harm to nonminority individuals (*Grutter* 2003, 343). Justices Ginsburg and Breyer expressed some reservations about Justice O'Connor's attempt to fix an end point to the underlying inequities that support the use of affirmative action admissions programs (344–346).

The concern about harming innocents is, in the end, what lies beneath the claims that affirmative action is reverse discrimination (Eastland 1996). According to this argument, affirmative action unavoidably benefits minority applicants at the expense of white applicants who have done "nothing wrong." As a result, these innocent white applicants are victims of reverse discrimination, because their only offense was being born white and their race is being used against them in an admissions decision and in a manner inconsistent with the equal protection clause of the Fourteenth Amendment.

Arguments against Affirmative Action: Backlash

Following hard upon the connections between harm to innocents and reverse discrimination is the argument that affirmative action programs lead to resentment, hostility, and a backlash in the white community (Edsall and Edsall 1991; Graham 1992, 206–219). Many white Americans perceive affirmative action as providing blacks with an unfair and unnecessary advantage in a competitive world and, consequently, as disadvantaging them.[8]

This resistance has led to a racial reconfiguration of party alliances and voting patterns in American politics. Southern whites, who had historically aligned with the Democratic Party and against the Republican Party (the "Party of Lincoln"), slowly began in the 1960s and 1970s to move away from the Democratic Party during the civil rights period (Graham 1992; Lawson 1976) . By the time of Ronald Reagan's presidency, Southern whites began to form a solid bloc of politically conservative Republican supporters, in large part due to their resistance to the policies (such as affirmative action) that were endorsed by the Democratic Party. White Americans who left the Democratic Party in these decades began to resent "the special status of blacks..." (Edsall and Edsall 1991, 182). For these individuals, affirmative action is a concrete manifestation of this objectionable special status afforded to blacks and other minority groups.

Arguments against Affirmative Action: Denying Individuality

Another argument against affirmative action is that it contradicts a fundamental principle of the U.S. constitutional tradition: individuals hold their rights as individuals, not as members of groups (*Bakke* 1978, 289–291). According to the individual rights perspective,[9] the fundamental problem with affirmative action is not (just) that it grants a social advantage to racial and ethnic minorities on the basis of their race; the problem is that, in doing so, affirmative action does not treat these individuals as individuals. Instead, affirmative action grants the same rights to all members of certain groups solely on the basis of their membership in those groups (Fried 1990, 108–109).

This argument does not insist that the principle of color-blindness prevents any consideration of race whatsoever when formulating public policy.[10] However, it requires some element of "victim-specificity" (111). In other words, there must be some demonstration that a particular individual has suffered a harm for which she is entitled to a legal remedy, before that remedy may be given (*City of Richmond* 1989, 526–528). On this view, simple membership in an historically disadvantaged group, without more, is insufficient to meet this individualized demonstration of legal harm.

Arguments against Affirmative Action: Undermining Achievement and Inducing Doubt

Affirmative action is especially unfair to minority applicants who "do not need it," because the existence of the program raises doubts about whether any member of the relevant minority group earned her place in the class entirely on her own merit. Importantly, these doubts are raised in the minds of both the non-beneficiaries of the program as well as the minority group members themselves (Steele 1990, 116).[11] In fact, these doubts appear especially debilitating for minority students given some evidence that the fears of some black students that they will not perform well on tests actually cause these students to perform more poorly on these tests than they otherwise would (Bowen and Bok 1998, 81–82).

This may indicate that affirmative action prevents people from enjoying their work and appreciating their success, because of the reaction of others that their success was not fully "earned" or their own internalized doubts in this direction (Cose 1993, 122–123). This argument generally concedes that elimination of affirmative action will reduce, perhaps drastically, the presence of blacks and other racial and ethnic minorities in the most prestigious educational institutions in the United States. But the counterbalance to this reality is that those minority students who do gain admission will not constantly be forced to labor under the shadow of doubt, of others and themselves, about the basis of their accomplishments.

Arguments For Affirmative Action: Educational Disparity and Disadvantage

Assessed quantitatively or anecdotally, proponents of affirmative action argue that there is no credible way to deny that the average black student faces formidable educational obstacles of various sorts.[12] From lack of funding and resources to a lack of positive reinforcement to active discouragement, many black schoolchildren are informed implicitly, explicitly, and institutionally that they are not expected to succeed academically (Cose 1993, 161–162; Massey and Denton 1993, 141–142). Moreover, these negative expectations have been inculcated and culturally assimilated to the point where certain minority groups (notably segments of the

African American community) equate success with whiteness and view academic achievement as abandonment of their ethnic or racial identity (Fordham and Ogbu 1986, 177; Jencks and Phillips 1998, 9). The resulting loss of "effort optimism" means that these students no longer believe that there is any benefit (for them) to working hard and attempting to succeed academically (Task Force 1997/1998, 96).

This view often begins with the recognition that the experience of African Americans is unlike that of any other ethnic or racial minority group in the United States (with the possible exception of Native Americans).[13] A history of slavery and legalized segregation has led to the current residential "hypersegregation" of American society (Massey and Denton 1993, 74–78). This radical residential segregation leads directly to (and proceeds directly from) pervasive poverty and attendant disadvantages in education, environmental conditions, health care, and employment opportunities. As a result of this reality, a principle of color-blindness as applied to issues of racial equality is inappropriate:

> A racially segregated society cannot be a race-blind society; as long as U.S. cities remain segregated—indeed, hypersegregated—the United States cannot claim to have equalized opportunities for both blacks and whites. In a segregated world, the deck is stacked against black socioeconomic progress, political empowerment, and full participation in the mainstream of American life. (148)

On this view, race can and should be permitted in admissions decisions because race can and should be considered in any area of public life, perhaps none more than education, where systematic and institutionalized inequities have for so long been imposed on blacks solely because of their race. In these circumstances, color-blindness is not a laudable neutral principle of constitutional doctrine. In these circumstances, color-blindness is blindness: to historical facts and to their current socioeconomic effects.

Arguments for Affirmative Action: Future Promise and Past Performance

This argument in favor of affirmative action was mentioned in passing by Justice Powell in *Bakke*, but not pursued because the parties did not address the issue. Undergraduate and graduate school admissions should be based on past performance not necessarily for its own sake, but rather as an indicator of promise of achievement in the future. Standardized test scores are relatively poor indicators of native intellect and potential for success

(Gould 1981, 9–29). These inaccuracies are accentuated by the disparate social, economic, and educational backgrounds and opportunities of the students whose ability is measured by these tests for purposes of higher education admissions decisions. Properly understood, then, affirmative action programs help to redress cultural bias or disadvantage in educational opportunity and standardized testing by attempting to assess intrinsic ability rather than purely quantitative measures of achievement.[14]

From this perspective, the standardized test scores accentuated in higher education admissions simply reinforce all of the educational inequities identified in the previous argument (Brown-Nagin 2005, 797; Sturm and Guinier 1996, 968). Rather than systematize and institutionalize these disparities further by overemphasizing "objective" factors, affirmative action permits admissions officers to evaluate a candidate's ability to succeed in the process of giving that candidate an opportunity to succeed. As *Bakke* indicates, this process requires the holistic evaluation of each candidate as an individual whose identity consists of more than grades and test scores (and race). The point here is not that grades and test scores do not matter at all; the point is that grades and scores are not all that matters.

Arguments for Affirmative Action: Diversity

In the affirmative action debate, the compelling interest in educational diversity is the logical extension of the legal principle articulated in *Brown v. Board of Education*.[15] In *Brown*, the Supreme Court held that public schools (and public education) in the United States could not be segregated on the basis of race. The diversity justification for affirmative action means that black students and white students learn by going to school with one another. By learning with one another they will learn from one another (*Bakke* 1978, 312 n. 48). *Brown, Bakke,* and *Grutter* all recognize that, for black and white students to learn from one another they must be given the opportunity to learn with one another.

Social science studies support this view. Patricia Gurin's research was presented to the lower courts in the *Grutter* case, and was cited extensively by the U.S. Court of Appeals for the Sixth Circuit in its opinion upholding the Law School admissions program. As summarized by the Sixth Circuit, Professor Gurin's research supports the conclusion that diversity benefits minority and nonminority students who encounter different perspectives at a time when they are beginning to define their own identity, to think more broadly about various social and political issues, and to prepare themselves for life in the pluralistic society of the United States (*Grutter* 2002, 759–762).

Arguments for Affirmative Action: Constitutional Logic and Original Positions

One of the most innovative, and least well-known, arguments in favor of affirmative action takes as a starting point the reality of racial imbalance in various areas of public life. African Americans and other minority groups are disproportionately underrepresented in government and in various professions. Indeed, Justice Brennan referred to and highlighted this point specifically in his *Bakke* dissent.[16] Ronald Fiscus (1992) also noticed this racial imbalance and he concluded that "there are only two assumptions possible here: that the races are equal at birth, or the contrary, that the races are not equal at birth" (25). The historical claim of racial disparities in intellect has of course been scientifically discredited and, as Fiscus points out, is prohibited by the Fourteenth Amendment and its assumption of inborn racial equality (25–26).

Fiscus then argues that if we imagine a perfectly nonracist society, from the perspective of the "original position of equality,"[17] we would (and should) expect to see perfectly proportional representation of all races in all professions and areas of public life. The ideal of the Reconstruction Amendments is the constitutional requirement that, no matter what real people may believe or do, the Constitution of the United States requires that people be treated as though they lived in a society entirely devoid of all "badges and incidents"[18] of racism:

> The validity of our argument does not depend on the actual likelihood of achieving such a perfectly nonracist society, but simply on the truth of the claim that distributive justice requires thinking in terms of complete nonracism. For that reason alone, we must stipulate that in our hypothetical society the color of one's skin has absolutely *no* effect on people whatsoever, as if they were in fact color-blind. The society described above is the sort of society that one would think everyone has a right to grow up in. In terms of the spirit of equal protection it is the ideal society, one where race truly is irrelevant in all aspects of life. Distributive justice, as it relates to race, can only be determined by conceiving of the complete eradication of racism, even if that should prove to be a distant or even idle hope in practice. (18)

From here, Fiscus argues that quota systems in affirmative action admissions plans are not unconstitutional, because they instantiate the proportional presence of minority (and nonminority) students in all of the educational institutions where, in the absence of racism, those minority students would be, anyway (19).

In the legal and policy debate about affirmative action in the United States, talk of "quotas" has become taboo.[19] Fiscus's argument is an effort to resuscitate the constitutional and theoretical bases for that argument. And it is worth noting that, while this view has never garnered a majority of the votes on the Supreme Court, it was accepted by four of the justices who decided *Bakke* (*Bakke* 1978, 378–379). Moreover, Fiscus's argument also supports the more general claim that affirmative action is, necessarily, a partial and artificial effort to correct for the racial imbalance and intolerance that have always existed in U.S. society. Affirmative action cannot eliminate the racism that has perpetuated the intolerance. But given that the intolerance will never disappear, affirmative action can at least correct the imbalance (West 2001, 95).

Arguments for Action: Civic Engagement and Social Responsibility

Another argument in favor of affirmative action is that it instills in its recipients a sense of community activity and contribution. Affirmative action engenders a sense of connection and responsibility to broader groups and coalitions that manifests itself in various ways. For example, data indicate that affirmative action increases the likelihood that individuals will volunteer time to social service organizations, participate in youth activities, join various religious congregations, arts institutions, and alumni associations (Bowen and Bok 1998, 155–160). In addition to participating in these various activities, affirmative action also increases the chances that individuals will be leaders in their communities and organizations (160–173). And affirmative action programs often encourage active interest and participation in politics and public life (173–174).

This heightened sense of connection and responsibility also translates into a strong sense of satisfaction and contentment on the part of minority graduates of these educational institutions. As I indicated earlier, one argument against affirmative action is that it increases self-doubt for some blacks and other minority group members. On this point, it is difficult to draw any definitive conclusion. Minority graduates of highly competitive educational institutions report solidly positive views of their own accomplishments and satisfaction with their lives (180–186). Yet these responses also tend to demonstrate lower trends of satisfaction and contentment than their white counterparts. It is difficult and unwise to extrapolate too broadly about the causes or meanings of this disparity (186–191). But it is noteworthy, nevertheless.

Conclusion

Tocqueville's comment about the judicialization of contentious political questions in the United States is at least as true today as it was when he wrote it. Surely, it is true of the affirmative action debate. Yet, as Cass Sunstein (1999) emphasizes, this is not necessarily a negative thing. Where the debate about affirmative action is concerned, the Court's decisions have helped both to indicate the need for extensive public debate about the issue and also to frame that debate. In this area, like many other divisive policy issues of its kind, "the Supreme Court can signal the existence of hard questions of political morality and public policy, by taking cases, drawing public attention to the underlying questions, and refusing to issue authoritative pronouncements" (131).

Bakke and *Grutter* are authoritative legal pronouncements, but they are best understood as the continuation of and participation in a larger national discussion. That larger debate needs the active, earnest, and respectful expression of all people who care about the process of deliberative democracy (132). The important point here is not to think of the debate over affirmative action in the United States as a debate about whether certain Supreme Court cases were decided correctly. The point is to see the cases as a contribution to the debate about what equality means and how it can best be realized in the United States. To achieve this understanding in a manner consistent with the best tradition of sustained public discourse in the United States, we must at least try to evaluate which arguments are worth considering most carefully.

Notes

1. By this statement, I do not mean to suggest that all concerted efforts to admit students of color to institutions of higher education began in 1961. On the contrary, several colleges and universities (notably selective liberal arts colleges) began to recruit minority students in the early nineteenth century, if not before (Duffy and Goldberg 1998).
2. As an illustration of this distinction, Justice Powell emphasized in *Bakke* that "rights created by the first section of the Fourteenth Amendment are, by its terms, guaranteed to the individual. The rights established are personal rights" (*Bakke* 1978, 289). Contrastingly, in China "the ultimate goal of educational policy with regard to national minorities is to achieve national integration. Maintaining cultural autonomy within a national framework is most often a struggle for minorities. States resist anything that leads to the disintegration of

national unity" (Postiglione 1992, 329). Of course, the concept of integration is itself contested politically and legally in both nations, as well.

3. For purposes of this discussion, I do not distinguish between constitutional challenges to affirmative action and statutory challenges under, e.g., Title VI of the Civil Rights Act of 1964. As they relate to affirmative action plans, the constitutional and statutory requirements are frequently read to coincide (*Bakke* 1978, 340, 352–353).
4. I should note that the converse is true for affirmative action programs in the employment context. In general, remedying specific instances of past discrimination—and not attempting broadly to increase diversity—is the basis upon which an affirmative action program may proceed in employment contracting, hiring, promotion, and so on. According to the Supreme Court decision in *Wygant v. Jackson Board of Education* (1986), "societal discrimination, without more, is too amorphous a basis for imposing a racially classified remedy... No one doubts that there has been serious racial discrimination in this country. But as the basis for imposing discriminatory *legal* remedies that work against innocent people, societal discrimination is insufficient and over-expansive. In the absence of particularized findings, a court could uphold remedies that are ageless in their reach into the past, and timeless in their ability to affect the future" (276).
5. Although it ultimately proved unsuccessful in *Bakke*, these four justices were attempting to extend the scope of two earlier decisions (*Califano* 1977, 317; *Craig* 1976, 197).
6. In a companion case involving the University of Michigan's undergraduate affirmative action program, the Court struck down the plan as unconstitutional. Unlike the Law School's affirmative action plan, the University of Michigan undergraduate admissions plan used a point system that granted minority applicants twenty points by virtue of their minority status. The Court held that the undergraduate plan, unlike the Law School's plus-factor plan, precluded individualized assessment and comparison of minority applicants with other applicants as required by Justice Powell's opinion in *Bakke* (*Gratz* 2003, 271–274). In contrast to the Supreme Court's *Gratz* decision, China employs an aggressive bonus point system that favors minority applicants. However, unlike the University of Michigan system struck down in *Gratz*, under the Chinese system bonus points are added to an applicant's entrance examination score, rather than to the individual's application file (Sautman 1999, 189).
7. The Court is quoting the *Bakke* decision here (*Bakke* 1978, 317).
8. There is evidence that similar resentments exist within the Han community in reaction to the positive action preferential admissions policies of China. These reactions seem far more muted than in the United States, however, because (among other reasons) preferential admissions policies are not perceived as interfering significantly with the educational options of Han students (Sautman 1999, 194).
9. I use this term to distinguish the individual rights perspective from the competing "group-rights perspective" (Fried 1990, 109).

10. Fried (1990) does not argue for absolute color-blindness as a constitutional principle: "It is impossible to ignore racial differences entirely—pure colorblindness is too extreme a principle" (111).
11. There is evidence that similar uncertainties and insecurities exist in China among minority students with respect to their Han contemporaries (Sautman 1999, 196).
12. Like the experience of blacks in the United States, although of course for different historical and cultural reasons, minorities in China generally experience significantly diminished educational achievement (Postiglione 1992, 315).
13. Justice Marshall offered a trenchant articulation of this perspective in his *Bakke* (1978) opinion: "It is unnecessary in 20th century America to have individual Negroes demonstrate that they have been victims of racial discrimination; the racism of our society has been so pervasive that none, regardless of wealth or position, has managed to escape its impact. The experience of Negroes in America has been different in kind, not just in degree, from that of other ethnic groups" (400).
14. According to this argument's proponents, affirmative action programs attempt to achieve "fair appraisal of each individual's academic promise in the light of some cultural bias in grading or testing procedures. To the extent that race and ethnic background were considered only to the extent of curing established inaccuracies in predicting academic performance, it might be argued that there is no 'preference' at all" (*Bakke* 1978, 306).
15. For example, David Strauss (1986) argues that "affirmative action is not at odds with the principle of nondiscrimination established by *Brown* but is instead logically continuous with that principle" (100).
16. Here is the pertinent language from Brennan's opinion: "In 1950, for example, while Negroes constituted 10% of the total population, Negro physicians constituted only 2.2% of the total number of physicians... By 1970, the gap between the proportion of Negroes in medicine and their proportion in the population had widened: The number of Negroes employed in medicine remained frozen at 2.2% while the Negro population had increased to 11.1%" (*Bakke* 1978, 369–370).
17. Fiscus borrows this concept from John Rawls (1971, 12).
18. This phrase comes from the *Civil Rights Cases*: "the Thirteenth Amendment... has a reflex character also, establishing and decreeing universal civil and political freedom throughout the United States; and it is assumed, that the power vested in Congress to enforce the article by appropriate legislation, clothes Congress with power to pass all laws necessary and proper for abolishing all badges and incidents of slavery in the United States..." (*Civil Rights Cases* 1883, 20).
19. By contrast, in China quotas have been used systematically to maintain minority enrollment at major universities (Sautman 1999, 185–189). Unlike Fiscus's rationale, however, quotas are not necessarily used to ensure proportional minority presence in higher education. In certain Chinese provinces, the minority presence engendered by quota implementation may over- or underrepresent certain minorities in relation to their percentage of the total population (Sautman 1999, 188).

References

Adarand Constructors, Inc. v. Pena, 515 U.S. 200 (1995).

Bowen, William G., and Derek Bok. 1998. *The Shape of the River: Long-Term Consequences of Considering Race in College and University Admissions*. Princeton, NJ: Princeton University Press.

Brown v. Board of Education, 347 U.S. 483 (1954).

Brown-Nagin, Tomiko. 2005. "The Transformative Racial Politics of Justice Thomas?: The *Grutter v. Bollinger* Opinion," *University of Pennsylvania Journal of Constitutional Law* 7: 787–807.

Califano v. Webster, 430 U.S. 313 (1977).

Carter, Stephen L. 1991. *Reflections of an Affirmative Action Baby*. New York: Basic Books.

City of Richmond v. J.A. Croson Company, 488 U.S. 469 (1989).

Civil Rights Cases, 109 U.S. 3 (1883).

Cose, Ellis. 1993. *The Rage of a Privileged Class*. New York: Harper Collins.

Craig v. Boren, 429 U.S. 190 (1976).

Duffy, Elizabeth A., and Idana Goldberg. 1998. *Crafting a Class: College Admissions and Financial Aid, 1955–1994*. Princeton, NJ: Princeton University Press.

Eastland, Terry. 1996. *Ending Affirmative Action: The Case for Colorblind Justice*. New York: Basic Books.

Edsall, Thomas Byrne, and Mary D. Edsall. 1991. *Chain Reaction: The Impact of Race, Rights and Taxes on American Politics*. New York: W.W. Norton.

Fiscus, Ronald J. 1992. *The Constitutional Logic of Affirmative Action*. Durham, NC: Duke University Press.

Fordham, Signithia, and John U. Ogbu. 1986. "Black Students' School Success: Coping with the Burden of 'Acting White,'" *Urban Review* 18: 176–206.

Fried, Charles. 1990. "*Metro Broadcasting, Inc. v. FCC*: Two Concepts of Equality," *Harvard Law Review* 104: 107–127.

Gould, Stephen Jay. 1981. *The Mismeasure of Man*. New York: W.W. Norton.

Graham, Hugh Davis. 1992. *Civil Rights and the Presidency: Race and Gender in American Politics, 1960–1972*. New York: Oxford University Press.

Gratz v. Bollinger, 539 U.S. 244 (2003).

Grutter v. Bollinger, 539 U.S. 306 (2003).

Grutter v. Bollinger, 288 F.3d 732 (6th Cir. 2002).

Hopwood v. Texas, 78 F.3d 932 (5th Cir. 1996).

Jencks, Christopher, and Meredith Phillips. 1998. *The Black-White Test Score Gap*. Washington, DC: Brookings Institution Press.

Kane, Thomas J. 1998. "Misconceptions in the Debate Over Affirmative Action in College Admissions," in *Chilling Admissions: The Affirmative Action Crisis and the Search for Alternatives*, eds. Gary Orfield and Edward Miller, 17–31. Cambridge, MA: Harvard Education Publishing Group.

Kennedy, John F. 1961. Executive Order 10,925.

Krieger, Linda Hamilton. 1998. "Civil Rights Perestroika: Intergroup Relations after Affirmative Action," *California Law Review* 86: 1251–1333.

Lawrence, Charles R. III, and Mari J. Matsuda. 1997. *We Won't Go Back: Making the Case for Affirmative Action*. Boston: Houghton Mifflin.

Lawson, Steven F. 1976. *Black Ballots: Voting Rights in the South, 1944–1969*. New York: Columbia University Press.

Lempert, Richard O., David L. Chambers, and Terry K. Adams. 2000. "Michigan's Minority Graduates in Practice: The River Runs Through Law School," *Law and Social Inquiry* 25: 395–505.

Massey, Douglas S., and Nancy A. Denton. 1993. *American Apartheid: Segregation and the Making of the Underclass*. Cambridge, MA: Harvard University Press.

Plessy v. Ferguson, 163 U.S. 537 (1896).

Postiglione, Gerard A. 1992. "The Implications of Modernization for the Education of China's National Minorities," in *Education and Modernization: The Chinese Experience*, ed. Ruth Hayhoe, 307–336. New York: Pergamon Press.

Rawls, John. 1971. *A Theory of Justice*. Cambridge, MA: Harvard University Press.

Regents of the University of California v. Bakke, 438 U.S. 265 (1978) .

Rubenfeld, Jed. 1997. "Affirmative Action," *Yale Law Journal* 107: 427–472.

Sautman, Barry. 1999. "Expanding Access to Higher Education for China's National Minorities: Policies of Preferential Admissions," in *China's National Minority Education: Culture, Schooling, and Development*, ed. Gerard A. Postiglione, 173–210. New York: Falmer Press.

Schnapper, Eric. 1985. "Affirmative Action and the Legislative History of the Fourteenth Amendment," *Virginia Law Review* 71: 753–798.

Smith v. University of Washington Law School, 233 F.3d 1188 (9th Cir. 2000).

Steele, Shelby. 1990. *The Content of our Character: A New Vision of Race in America* New York: St. Martin's Press.

Strauss, David A. 1986. "The Myth of Colorblindness," *Supreme Court Review*: 99–134.

Sturm, Susan, and Lani Guinier. 1996. "The Future of Affirmative Action: Reclaiming the Innovative Ideal," *California Law Review* 84: 953–1036.

Sunstein, Cass R. 1999. *One Case at a Time: Judicial Minimalism on the Supreme Court*. Cambridge, MA: Harvard University Press.

Task Force Report of the American Psychological Association for 1996. 1997/1998. "Explaining the Gap in Black-White Scores on IQ and College Admissions Tests," *Journal of Blacks in Higher Education* 18: 94–97.

Thernstrom, Stephan, and Abigail Thernstrom. 2002. "Does Your 'Merit' Depend upon Your Race? A Rejoinder to Bowen and Bok," in *The Affirmative Action Debate* (second edn), ed. Steven M. Cahn, 183–189. New York: Routledge.

Tocqueville, Alexis de. 1945. *Democracy in America*. Ed. Phillips Bradley. New York: Alfred A. Knopf.

West, Cornel. 2001. *Race Matters* (second edn). New York: Vintage Books.

Wygant v. Jackson Board of Education, 476 U.S. 267 (1986).

Chapter 14

Native and Nation: Assimilation and the State in China and the U.S.

Ann Maxwell Hill

Native Americans and peoples in China designated "minorities" or "minority nationalities" would seem to have much in common if viewed through the contemporary lens of popular culture prevalent in the two nations. Both Native Americans and China's minorities have been described as "ethnic" peoples, groups outside the nation's majority population that represents the nation's culture and history. Ethnic groups are often seen as "tribal," a term with connotations of primitiveness and the implication that they have recently emerged from loosely organized tribal life, or otherwise failed to achieve the familiar political configurations of state or empire. Relative to the majority populations, these minorities may be romanticized as repositories of "traditional" knowledge and behaviors reflecting a closeness with nature now lost to the majority freighted with the knowledge of science and the experience of technology and urban life. The flip side of the romance with peoples uncorrupted by contemporary urban life is the ethnocentric notion that minorities need tutelage or direction from the majority population, a response to the related perception that they lag behind the times and are beset with problems arising from their inability to adapt to modern life. Such perceived shortcomings are often attributed to their tendency to cling to irrational customs, their geographic isolation (hence, their failure to learn the national language and culture), and general ideologies, more or less determinist, about their inferiority.

Behind these shared clichés, however, are a host of differences between American Indians and minorities in the People's Republic of China (PRC) and in the way they are positioned within, and sometimes outside of, the nation. Focusing on assimilation, I use a broad comparative frame for examining some of these differences, largely historical and cultural, that are often taken for granted in policy discussions. The time frame, too, is broad—roughly the late nineteenth century through the twentieth century. In the case of China, the time period facilitates consideration of the late imperial era and its inevitable legacy for twentieth-century nationalists, just as in North America, a focus that begins with the late nineteenth century captures the coercion of forced assimilation and expropriation of American Indians that continued to reverberate into the twentieth century.

In the spirit of this volume, which is intended to reach a wide audience and, in particular, to open a dialogue between those in the United States and China with vested interests in affirmative action in minority education, I point out contrasts between the two national cases that may help the reader unfamiliar with one nation or the other, or both, to better understand the sometimes disparate contexts of contemporary affirmative action in education globally. One might well ask why, in this comparative framework, Native Americans are the subjects of comparisons with China's minorities, rather than African Americans, whose proportion in the U.S. population (about 13 percent versus 1.5 for Native Americans) is larger and whose impact on affirmative action far greater. The answer is simple: there are no minorities in China that have experienced institutionalized slavery on a scale comparable to that of African Americans. As my opening comments on shared stereotypes illustrates, Native Americans as indigenous people have much more in common with the historical experience of China's minorities than do African Americans.

From the vantage point provided by comparison, assimilation as an ideal dominating popular discourse and state policies in the United States stands out as the main tool intended to bring native peoples into the nation. The effectiveness of assimilationist projects, whether in colonial America or much later under the twentieth century's failed reservation system, is another matter. In fact, and here speaking generally of the much maligned "melting pot" view of the mid-twentieth century, it was only when assimilation was widely acknowledged as a failure that race-based affirmative action was found appealing.

While the legacy of empire is not completely without precedent in U.S. colonial history—the British, after all, brought with them from Ireland their own imperialist baggage—the process of incorporation of non-Han peoples into the contending national projects in China's twentieth century in part reflected the perennial dilemmas of empire and its far-flung frontiers.

Viewed from the center, the strategic location of many of China's native peoples on the historical borders of empire required political allegiance first and foremost, rather than assimilation. This is not to say that assimilation as an agenda was absent from concerns of the Qing Dynasty (1644–1911) for border management and, later, from the process of nation-building in Republican-era China (1912–49) and under the PRC government. Rather, I mean that China's nation-making, because of the legacy of powerful and strategically located unassimilated peoples on her frontiers, entailed many more political accommodations and necessarily a much more pluralist outlook than in North America. Nor am I gainsaying the later impact of the Soviet model on China's revolutionary leaders as a blueprint for bringing nationalities into the socialist polity (see chapter two). The Soviet model, too, had to accommodate some of the same legacies of empire as did those of China's nationalists.

The Marginality of Indigenous Peoples?

The marginality of indigenous peoples is a condition familiar to anthropologists and certainly a staple of colonial histories in the Americas. However, indigenous people's marginality, as a concept positioning them vis-à-vis the state and one that connotes powerlessness and tribalism, does not travel well (Bodley 1999, 4). Moving back in time, and outward from China's center to the edges of empire, one encounters powerful state-organized societies that sometimes overtook the central Chinese state and even in the twentieth century presented daunting challenges to nationalist unifiers. Moreover, in the face of the indigenous origins of Chinese populations themselves, homegrown, so to speak, what serves to contrast majority populations from Native Americans in the United States is of limited utility in the Chinese case.

The inappropriateness of the trope of the marginality of indigenous peoples as universal signals more than a terminological issue. Marginalization of American Indians began with depopulation under the onslaught of European-born diseases, continued with the various Indian removal schemes epitomized by Andrew Jackson's policies in the early nineteenth century, and culminated in pressures from settlers and missionaries for Indian lands and converts. Most Indians in colonial and later U.S. territories lived in comparatively loosely organized, scattered communities unable to repel state agents armed with superior technology, and settlers, whose sheer numbers were overwhelming. Of the late nineteenth and early twentieth centuries, in particular, one might well say that

Native Americans were ultimately undone by the plow, rather than the sword.

Minority populations in China, by contrast, included groups with state traditions, as well as small-scale kin-based groups conforming more closely to our notion of indigenous peoples. Among state peoples, the Dai of southwest China were Buddhist, literate, and closely allied with even larger states to the south and west; in their relationship with the Qing and Republican-era governments, they were treated often more like vassals than subject peoples. In the northwest, there were populations more or less integrated into Mongol and Muslim feudal polities that throughout the Qing mounted substantial challenges to the empire's elites on both the periphery and in Beijing, pressures for local autonomy that continued under various warlords during the Republican era. The Qing rulers themselves were Manchu, with roots in northern frontier populations. Most Western historians of China, sometimes at odds with interpretations of their Chinese colleagues on this matter, acknowledge that Qing authority over peoples at the periphery of empire "cannot be described as monolithic or consistent" (Lipman 1997, xxix), and current scholarship on Nationalist China (1925–49) demonstrates that the regime's control over ethnic border regions was intermittent and precarious, their commitment to restoring the integrity of China's territorial and sovereign rights after the initial failures of the new republic notwithstanding (Lin 2006, 12). Especially in the southwest, rugged terrain and special adaptations in agriculture and herding enabled non-Chinese societies to occupy ecological niches unattractive to Chinese settlers and inaccessible to state armies.

Depopulation

Another important difference between the relative marginality of Native Americans and China's ethnic minorities has been the proportion of their population in the context of the total population of the nation. Assimilation as a proposition for dealing with people beyond the majority seemed a practicable goal to white Americans and their government after the defeat of the last remnants of Native American warriors at the turn of the twentieth century, the nadir of the decline of Native American populations.

European diseases and common viruses, unwittingly transmitted to Native Americans, had devastating effects. While there is general agreement that depopulation among native peoples of North America was substantially affected by their lack of immunities to Old World diseases, recent research has also demonstrated that "the indirect effects of disease episodes appear more important in population decline" than mortality from European-born diseases per se (Thornton 1997, 311). This new interpretation would account for the fact that population declines among Native Americans continued throughout the seventeenth, eighteenth, and nineteenth centuries, reaching their lowest point as late as 1900. Increased mortality and decreased fertility, goes the newer argument, were also related to colonial and colonialist practices, such as forced removal and concomitant stress and starvation.

After 1900, Native American populations slowly began to recover. From an estimated total of 250,000 at the end of the nineteenth century, the Native American population grew at a rate of about 7 percent per decade, reaching 380,000 by 1950 (Shumway and Jackson 1995, 186). Probably some of the growth was due to better surveys: Native Americans, deemed unfit for citizenship in the nineteenth century, were granted that status in 1924. The change in status meant that Native Americans living off the reservations were, for the first time, counted in the U.S. census. The latest census data from 2000 indicate that today there are 4.3 million Native Americans and Alaska Natives, about 1.5 percent of the total U.S. population. Only about a third of this number are Native Americans with official tribal registrations (Ogunwole 2006).

One of the natural consequences of centuries' long contact between Chinese-identified populations and those in adjacent territories was the mutual adaptation to local environments and development of immunities to local viruses and some epidemic diseases. In other words, unlike in colonial North America, diseases characteristic of the frontier zones in China during the nineteenth and twentieth centuries did not result in rapid depopulation of native peoples, as migrants and soldiers from China's interior made their way to more remote areas. On the contrary, the perception among Chinese officials, Qing and Republican, was that diseases endemic to Yunnan's major river valleys were lethal to Han Chinese, rather than the other way around. This belief about the greater susceptibility of Han to malarial diseases as compared to local peoples may or may not be borne out by contemporary science, but in fact both populations, local and migrant, suffered from malaria, a disease transmitted by mosquitoes rather than movements of people (Bello 2005). The consequences were more administrative than demographic—malarial districts were thought to require native rulers rather than Han bureaucrats.

Enumerating frontier peoples ancestral to China's minorities has proven quite difficult, for the simple reason that they often lived in territories separate from the local Han and administered by native leaders. They were "illegible" to the state, in James Scott's felicitous term, and therefore not readily available as taxpayers and providers of corvee labor (Scott 1998, 78).

Chinese land and tax registers in the Qing, for example, "were a function of territory, not, as has been thought, of ethnicity" (Lee 1982, 724). So land and tax registers are not necessarily helpful in sorting out local Han from local non-Han, and in any case it is well to keep in mind that Han living on the empire's, and later the nation's, frontiers were themselves ethnically mixed in terms of language, culture, and native place. In the Qing, in particular, and continuing under the Nationalists, migrants from China's interior flocked to frontier areas. This modern complex of migratory movements is not yet well understood, but my point here is that there is plenty of evidence that in some frontier areas, the influx of migrants (usually identified as Han) from the interior displaced local indigenous people, or, alternatively, created tensions and violent conflicts over resources along the frontiers. The presence of Han migrants—their often overweening role in local autonomous minority governments and their impact on minority cultures and economies—is a provocative issue today in many minority autonomous areas, but especially in Tibet and Muslim Xinjiang (Millward 2007, 348–352; Sautman 2006).

Because statistics on historical demography are in short supply, it may help to keep in mind that currently ethnic minorities occupy about five-eighths of China's total territory, including most of the strategically important border areas, a reflection of their territorial extent historically (Mackerras 2003, 15). Beginning with the founding of the PRC in 1949, better statistics were collected. Official records indicate that China's minority population grew from roughly 34 million in 1953 to its current size of 106.5 million, an increase accompanied by a rise in the minority populations' proportion in China's total population, from 5.9 percent to today's 8.4 percent. This is a stark contrast, both in terms of size of home territories and of proportion in the national population, to the numerical strength and miniscule reservation lands of Native Americans.

Civilizing Missions

The civilizing mission of the imperialist state, a staple concern of subaltern and postcolonial studies (e.g., Mamdani 2002; Said 1979), has also been implicated in relations between the center and peripheries of modern China of the eighteenth, nineteenth, and twentieth centuries. Re-popularized by Harrell (1995) and reconsidered in the plethora of new works among Western historians focused on issues of governance and culture at China's ethnic frontiers (e.g., Atwill 2005; Crossley et al. 2006; Elliot 2006; Giersch 2001; Herman 1997, 2007; Rowe 1994), civilizing missions in general were aimed at peoples defined as inferior to the civilizing agents, yet capable of improvement under the aegis of the civilizers (Harrell 1995, 8).

The question for this comparative chapter, especially looking at the "mission" more literally and effectively undertaken in the name of Christianity in North America, is whether or not the historic moral purpose of Confucianism to enlighten non-Chinese peoples through good government and education was synonymous with an assimilationist agenda. A related question is the more pragmatic one of the effectiveness of the Mencian idea of *hua*, or moral transformation, deployed intermittently by the Qing court in policies aimed at non-Chinese populations.

The first question about the discourse on assimilation and the Manchu/Qing government agenda in the southwest can only be answered approximately. Crossley et al. (2006), after pointing out that the Qing rhetoric of civilizing was less ardent and less consistent than in the Ming Dynasty (1368–1644), also acknowledged the strong Confucian moral tone of transforming the people but warned against conflating it with acculturation or assimilation: "This was the rhetoric of moral transformation, not of acculturation alone. If we are less than precise on this issue, it could appear at a casual reading that this expansionist rhetoric constituted an early modern theory of what is sometimes called 'sinicization'" (9). The sinicization model, with its connotations of automatic assimilation of non-Han peoples to Chinese society and polity, was not what the Qing bureaucrats always had in mind. And there is general agreement among contemporary historians of ethnicity in China that during the Qing and Republican eras, China's ethnic borders, whether in the interior or on the periphery, were areas of negotiation and sites of local agency on the part of minority populations. Acculturation, or the selective, expedient adoption of some Chinese cultural traditions, by local frontier peoples was a more likely scenario than assimilation (Giersch 2006, 188; Harrell 2001, 263). However, both assimilation and acculturation were long-term, messy processes, with cultural elements and identities crossing borders in multiple directions, including assimilation of Han settlers to local cultures (Fang 2003, 728, for examples from Guizhou).

If the Qing and Republican governments had been truly preoccupied with assimilation as a tool of political consolidation, then the historical record would be replete with education projects, since education was rhetorically significant as a source of transformation to both the Qing neo-Confucianists and early twentieth century reformers and revolutionaries. In fact, education efforts under the Manchus directed at civilizing various peoples on the borders of Chinese-occupied areas, although vigorously promoted by the Yongzheng (r. 1723–35) official Chen Hongmou, were short-lived (Rowe 1994, 443–444). In the same time period, the order of the day was military conquest of the frontier; native officials in particular were to be eliminated rather than assimilated (Giersch 2006, 44–45). In the Republican period, the Nanjing (Nationalist) government saw schools as opportunities to educate students in the Nationalist Party philosophy and to inculcate a sense of loyalty to the nation—a modernist civilizing project in its own right but one directed primarily at Chinese (Wu 2002, 171). Educators in the 1930s saw these goals as appropriate to projects for minority schooling, but exhibited a much keener sense of skepticism about outcomes than did their Qing counterparts: "Ethnographers, educators, and policymakers associated with these latter-day efforts despaired that Yunnan aborigines still spoke almost no Chinese, held fast to tribal identities, and violently resisted government attempts to alter their life-styles. Such men categorically dismissed all earlier programs of acculturation through education as utter failures" (Rowe 1994, 445).

As many U.S. historians have noted, a sense of mission—whether expressed in the religious terms of the early colonists, in the later nineteenth-century doctrine of Manifest Destiny, or in twentieth-century boarding schools for Indian children—was endemic in the response of European-descended settlers to Native Americans. "From its inception, the invasion of North America was launched on waves of pious intent," and discourses on the conversion of Native Americans to Christianity were a staple of colonial and later national policies (Axtell 2001, 146). Surprisingly, perhaps, the conversion process was always less about souls and more about making Native Americans into lesser versions of white Europeans, in hygiene and other bodily regimens, livelihood, sex roles, diet, and so on (Lomawaima 1993).

In 1879, the Carlisle Indian Industrial School was set up in former army barracks in Carlisle, Pennsylvania. The Indian School was the result of a proposal to Congress by Richard Pratt, an army officer and reformer who had long served in Indian territories and was sympathetic to Native Americans. He was particularly outraged by their inadequate food rations on reservations and the degrading treatment of captured warriors. Like most men of his time and status, he was religious, although by no means a missionary. If the civilizing metaphors of his plan were Christian, his explicit goal was total assimilation, to make the Indian into "a copy of his God-fearing, soil-tilling, white brother" (Landis 1996). By 1900, 85 percent of Indian children who were attending school were enrolled in mostly government-run boarding schools, either on or off the reservation. Although the Carlisle Indian Industrial School was closed in 1918, boarding schools remained the mainstay of Indian education until the 1970s and for Native Americans are forever associated with forced assimilation.

Roughly coincident with the founding of boarding schools and reservations was the Dawes Act passed in 1887. Preceded by a long history of pressure from white settlers and speculators for land, the Dawes Act was the logical culmination of the popular doctrine of Manifest Destiny, a complex of ideas about the right of Americans to extend their territory to the Pacific. The act reflected a mixture of specific motives ranging from principled opposition to the reservation system because it impeded full assimilation of Native Americans, to the view that the Indians had too much land that otherwise was necessary for economic growth and private development. The assimilationist agenda was to be found in the idea that by giving Indians title to a specific allotment of land, it would encourage the Indian family "to farm its own land and, in so doing, acquire the habits of thrift, industry and individualism needed for assimilation into white society" (Carlson 1978, 274). "Excess" land would be sold off. The Dawes Act was repealed in 1934 because it was widely recognized as ineffective in promoting family farming among Native Americans. However, it has been judged much more successful in its goal of placing reservation lands in the hands of white settlers (274).

Calculated Differences

Because of the variable but persistent strains of Christian doctrine in historical perceptions of Native Americans, racism as we know it today was slow to develop. Christian theories of the monogenesis of humans, all descended from a common ancestor, and the Christian desire to see Indians as amenable to conversion, compelled a view of European-native differences as less important than the human commonalities between the two groups (Sturm 2002, 44–45). However, as white settlers pushed west in the eighteenth and nineteenth centuries, conflict with Indians increased, and embittered whites were less likely than the seventeenth-century English colonists to acknowledge the humanity of their proximate enemies. The spread of plantation agriculture also contributed to the dissemination of racist ideologies. As Ira Berlin (1998), a well-known historian of American slavery, remarked:

> Drawing power from the metropolitan state, planters—who preferred the designation "masters"—transformed the societies with slaves of mainland North America into slave societies. In the process, they re-defined the meaning of race, investing pigment—both black and white—with a far greater weight in defining status than heretofore. (96)

The science of the time also played a role in the varieties of racial invention in the nineteenth century, when a burgeoning American race-based ethnology challenged biblical theories of a single origin for humankind (McLoughlin and Conser 1989, 250–251). Measuring skulls and assessing skin color, scientists and would-be scientists reddened the Indians. The racialized interpretations of difference were so pervasive in the nineteenth century, and at times useful to Indians with their own designs on nation, that the colors of race made their way into Native American origin myths (258–259).

The "science" of anthropology born in the 1920s was in part a response to the anti-immigrant eugenics movement at the time. Henceforth, American anthropologists argued against race as the sine qua non of human difference, substituting in its place culture. Although the work of cultural anthropologists was much more widely read in the first half of the twentieth century than it is today, it seemed to have made little headway in the popular culture of the United States, where segregation continued to feed on constructions of black-white racial differences. Meanwhile, biological anthropologists and paleontologists persisted in their search for race and racial origins in the fossil record, and the U.S. census, as well as more localized statutes of states and even tribal councils, identified Native Americans as a race, often relying on the idiom of "blood" and its quantity (based on the ethnicity of parents and earlier ancestors) to determine who was and was not Native American.

Until recently, American intellectuals have attributed race and racism to Western ideologies, especially those formed during the colonial era. Other scholars have pointed to the modern nation-state as the global political form most deeply implicated in raising race and ethnicity to the level of enabling domination of a majority population over minorities, who are assigned visible traits of racial and ethnic otherness (Williams 1989). To some extent, studies by Dikotter (1997) and others of racial formations in Asia have borne this out, as evidence for race in ideas about *minzu* and similar terms for descent-based notions of peoples came into the late-nineteenth-century discourse about nation, national unity, and culture (2–4). Moreover, as in the United States, racialized differences in Republican China were mixed up with enthusiasm for science and progress thought to be essential to the modern nation.

But for most anthropologists who study minorities in China, race is seen to have played a more muted role in conceptualizing difference than its centrality in the United States would predict. Although there is evidence for racial thinking in some Qing-era documents, biogenetic race was never an important metaphor for distinguishing Chinese from local populations on China's frontiers. In fact, labeling these populations as *shu,* or relatively civilized (literally, "cooked") or *sheng*, barbarian (literally, "raw"), at the very least signaled a predisposition toward acceptance of a continuum of distance from the Han, rather than an unbridgeable gulf.

Native and Nation

By the middle of the last century, Native Americans in the United States were largely out of sight, visible only on the screen in popular movies about the old West in Saturday matinees that resurrected the image of the noble savage. In the 1960s, energized by the Civil Rights Movement and the prospect of government dismantling of the reservations, Native Americans formed several pan-Indian organizations to gain recognition as a voice in national level decision-making. While some Native Americans, and not a few whites, dismissed these organizations as a return to conservative tribalism, others saw them as the inevitable outcome of failed "melting pot" policies and the inexorable descent of impoverished Indians into second-class citizenship (Hanson 1997). "Going back to the blanket," a slogan of the American Indian Movement, left no doubt as to how young Indian leaders saw their prospects for assimilation into mainstream U.S. society.

The pan-Indian movement also brought more scrutiny of the federal agency, the Bureau of Indian Affairs (BIA), historically charged with the management of all Indian affairs. From its creation in the first half of the nineteenth century to its reform in the 1970s, the BIA was heavily involved in almost every aspect of reservation life, and its mismanagement of Indian communities was widely known. The reforms of the 1970s committed the BIA to support tribal governments and improve reservation life. Beyond these platitudes has been the fact of the BIA's gradual withdrawal from direct involvement in Indian life and movement toward an advisory role.

By comparison, China at mid-century was confronted with the urgency of national consolidation after several decades of intermittent civil war, not to mention the Japanese occupation. As I have pointed out, minority nationalities in China occupied over half of the new nation's territory, including critical border areas that in the Qing and Republican eras were relatively autonomous. After the movement of the People's Liberation Army (PLA) into border areas, where they often met violent resistance, the allegiance of minority populations—roughly 6 percent of China's population in the 1950s—to the new Communist government of the PRC was secured through a variety of means: the promise of a degree of autonomy to be guaranteed by the constitution, delayed implementation of land reform policies in more remote areas, rewards to minority elites in the form of education and political participation in Beijing, policies that later on provided government services to minority areas while exempting them from some of the more Draconian policies imposed on the Han, most importantly, the one-child policy, and so on.

None of these measures was explicitly aimed at assimilation of the minorities, although the ten-year period of the Cultural Revolution (1966–76) pushed the Maoist agenda to its extreme and resulted in attacks on "superstition" and other "feudal" practices in minority areas. Minority education, a critical sphere for ideological and nationalist indoctrination in the PRC, has indeed raised literacy rates in Chinese among minorities, but has also given rise in the last two decades to an array of institutional arrangements at all levels of education for minority populations that facilitate bilingual programs, ethnic arts, and the training of minority cadres in ethnic histories and languages. Yet many Western researchers, most notably Harrell (1995) and Schein (2000), have pointed out that however much diversity or pluralism is promoted in theory, the state's hegemonic intellectual projects in education, ethnology, the arts, and other areas have defined contemporary minority identities, linking them inextricably to state ideologies that privilege the Han majority as exemplars of modernity and citizenship. The very process of ethnic identification described in the Introduction to this volume demonstrates the link between naming and governing that has been a persistent tool of national hegemony. Recent policies have moved away from the idea of a multinational nation, to a nation that is merely multiethnic, with a common language and common identity (see Introduction). In this light, China's positive policies toward minorities over the long haul may result in rates of assimilation to the Han majority higher than ever before, as more minority students are integrated into China's mainstream educational institutions.

References

Atwill, David. 2005. *The Chinese Sultanate: Islam, Ethnicity, and the Panthay Rebellion in Southwest China, 1856–1873*. Stanford: Stanford University Press.

Axtell, James. 2001. *Natives and Newcomers: The Cultural Origins of North America*. New York: Oxford University Press.

Bello, David. 2005. "To Go Where No Han Could Go for Long: Malaria and the Qing Construction of Ethnic Administrative Space in Frontier Yunnan," *Modern China* 31 (3): 283–317.

Berlin, Ira. 1998. *Many Thousands Gone: The First Two Centuries of Slavery in North America*. Cambridge: Belknap Press.

Bodley, John H. 1999. *Victims of Progress*. Mountain View, CA: Mayfield.

Carlson, Leonard A. 1978. "The Dawes Act and the Decline of Indian Farming," *The Journal of Economic History* 38 (1): 274–276.

Crossley, Pamela, Helen F. Siu, and Donald Sutton. 2006. "Introduction," in *Empire at the Margins*, eds. Pamela K. Crossley, Helen F. Siu, and Donald Sutton, 1–24. Berkeley: University of California Press.

Dikotter, Frank. 1997. "Introduction," in *The Construction of Racial Identities in China And Japan*, ed. Frank Dikotter, 1–11. Honolulu: University of Hawaii Press.

Elliott, Mark C. 2006. "Ethnicity in the Qing Eight Banners," in *Empire at the Margins*, eds. Pamela K. Crossley, Helen F. Siu, and Donald Sutton, 27–57. Berkeley: University of California Press.

Fang, Tie. 2003. *Xinan tongshi* [A Comprehensive History of the Southwest]. Zhengzhou: Zhongzhou guji chubanshe.

Giersch, C. Pat. 2001. "'A Motley Throng': Social Change on Southwest China's Early Modern Frontier, 1700–1880," *Journal of Asian Studies* 60 (1): 67–94.

———. 2006. *Asian Borderlands: The Transformation of Qing China's Yunnan Frontier*. Cambridge: Harvard University Press.

Hanson, Jeffrey R. 1997. "Ethnicity and the Looking Glass: The Dialectics of National Indian Identity," *American Indian Quarterly* 21 (2): 195–209.

Harrell, Stevan. 1995. "Introduction: Civilizing Projects and the Reaction to Them," in *Cultural Encounters on China's Ethnic Frontiers*, ed. Stevan Harrell, 3–36. Seattle: University of Washington Press.

———. 2001. *Ways of Being Ethnic in Southwest China*. Seattle: University of Washington Press.

Herman, John E. 1997. "Empire in the Southwest: Early Qing Reforms to the Native Chieftain System," *Journal of Asian Studies* 56 (1): 47–74.

———. 2007. *Amid the Clouds and Mist: China's Colonization of Guizhou, 1200–1700*. Cambridge, MA: Harvard East Asian Monographs 293.

Landis, Barbara. 1996. *Carlisle Indian Industrial School History*. Available online at http://home.epix.net/~landis/histry.html.

Lee, James Z. 1982. "Food Supply and Population Growth in Southwest China, 1250–1850," *Journal of Asian Studies* 41 (4): 711–746.

Lin, Hsiao-Ting. 2006. *Tibet and Nationalist China's Frontier: Intrigues and Ethnopolitics, 1928–49*. Vancouver: University of British Columbia Press.

Lipman, Jonathan N. 1997. *Familiar Strangers: A History of Muslims in Southwest China*. Seattle: University of Washington Press.

Lomawaima, K. Tsianina. 1993. "Domesticity in the Federal Indian Schools: The Power of Authority over Mind and Body," *American Ethnologist* 20 (2): 227–240.

Mackerras, Colin. 2003. "Ethnic Minorities in China," in *Ethnicity in Asia*, ed. Colin Mackerras, 15–47. London: Routledge Curzon.

Mamdani, Mahmood. 2002. "Good Muslim, Bad Muslim: A Political Perspective on Culture and Terrorism," *American Anthropologist* 104 (3): 766–775.

McLoughlin, William G., and Walter H. Conser. 1989. "'The First Man Red'—Cherokee Responses to the Debate Over Indian Origins, 1760–1860," *American Quarterly* 41 (2): 243–264.

Millward, James A. 2007. *Eurasian Crossroads: A History of Xinjiang*. New York: Columbia University Press.

Ogunwole, Stella. 2006. "We the People: American Indians and Alaska Natives in the U.S." http://www.census.gov/population/www/socdemo/race/censr-28.pdf.

Rowe, William T. 1994. "Education and Empire in Southwest China: Ch'en Hung-mou in Yunnan, 1733–38," in *Education and Society in Late Imperial China, 1600–1900*, eds. Benjamin E. Elman and Alexander Woodside, 417–457. Berkeley: University of California Press.

Said, Edward. 1979. *Orientalism*. New York: Vintage.

Sautman, Barry. 2006. "Tibet and the (Mis-) Representation of Cultural Genocide," in *Cultural Genocide and Asian State Peripheries*, ed. Barry Sautman, 165–272. New York: Palgrave MacMillan.

Schein, Louisa. 2000. *Minority Rules: The Miao and the Feminine in China's Cultural Politics*. Durham: Duke University Press.

Scott, James C. 1998. *Seeing Like a State: How Certain Schemes to Improve the Human Condition Have Failed*. New Haven: Yale University Press.

Shumway, Matthew J., and Richard H. Jackson. 1995. "Native American Population Patterns," *Geographical Review* 85 (2): 185–201.

Sturm, Circe. 2002. *Blood Politics: Race, Culture, and Identity in the Cherokee Nation of Oklahoma*. Berkeley: University of Oklahoma Press.

Thornton, Russell. 1997. "Aboriginal North American Population and Rates of Decline, ca. A.D. 1500–1900," *Current Anthropology* 38 (2): 310–315.

Williams, Brackette. 1989. "A Class Act: The Race to Nation Across Ethnic Terrain," in *Annual Reviews in Anthropology*, vol. 18, eds. Bernard J. Siegel, Alan R. Beals, and Stephen A. Tyler, 401–444. Palo Alto: Annual Reviews, Inc.

Wu, David Yen-Ho. 2002. "The Construction of Chinese and Non-Chinese Identities," in *China Off-Center: Mapping the Margins of the Middle Kingdom*, eds. Susan Blum and Lionel Jensen, 167–182. Honolulu: University of Hawaii Press.

Contributors

Zhanlong Ba, a Yugur native, earned his PhD in ethnology at the Central University for Nationalities in Beijing in 2008. He is currently a lecturer in the College of Social Development and Public Policy at Beijing Normal University. His research covers educational anthropology, developmental anthropology, ethnology, and social sciences. He has published over twenty articles on these topics.

Christina Y. Chan received her bachelor's degree in economics from the University of Washington in 2007. She participated in the UW Worldwide Sichuan University Exchange in 2005. Christina is currently a 2011 candidate for Juris Doctor at Northwestern University School of Law.

Walker Connor is currently distinguished visiting professor of political science at Middlebury College. He has held resident appointments at, inter alia, Harvard, Dartmouth, the London School of Economics, the Woodrow Wilson International Center for Scholars, Oxford, Cambridge, Bellagio, Warsaw, Singapore, Budapest, and Queen's University in Kingston. The University of Nevada named him Distinguished American Humanist of 1991/92 and the University of Vermont named him the Distinguished American Political Scientist of 1997. He has published over sixty articles and five books dealing with the comparative study of nationalism.

Yanchun Dai serves as an editor in the Yunnan Provincial Office of Local History and is a doctoral candidate at Yunnan University. Her research is focused on economic anthropology, with a specialty in the Miao people in Guizhou.

Douglas E. Edlin is an associate professor in the Department of Political Science at Dickinson College. He holds a JD from Cornell Law School and a PhD in public law from Oxford University. He is the author of *Judges and Unjust Laws: Common Law Constitutionalism and the Foundations of Judicial Review* (University of Michigan Press, 2008) and the editor of *Common Law Theory* (Cambridge University Press, 2007).

Gelek, born in eastern Tibet (Kham), earned his MA in ethnology and history of China's minority nationalities at the Graduate School of the Chinese Academy of Social Sciences in 1981 and his PhD in sociocultural anthropology at Zhongshan University in Guangzhou in 1986. He is the first Tibetan to earn a PhD in China. Since 1981, he has frequently worked with the Chinese central and local governments, international organizations, academic institutions, and private foundations. He has published thirteen scholarly books and over eighty articles. Currently he is professor and deputy director of the China National Center for Tibetan Studies in Beijing.

Stevan Harrell is professor of anthropology at the University of Washington, where he directs the UW Worldwide Sichuan University Exchange. His current research deals with the traditional ecological knowledge of the Nuosu people and the potential for building socially and ecologically resilient communities. He is also president of the Cool Mountain Education Fund, a small nonprofit organization that gives scholarships to middle- and high-school students from Yanyuan County, Sichuan.

Ann Maxwell Hill is professor of anthropology at Dickinson College. Her earlier fieldwork was conducted in Northern Thailand (*Merchants and Migrants: Ethnicity and Trade Among Yunnanese Chinese in Southeast Asia*, Yale SE Asia Monograph 47, 1998). For the past decade, she had done fieldwork in Nuosu communities in southwest China, the basis for several scholarly articles.

Evelyn Hu-DeHart is professor of history and ethnic studies, and director of the Center for the Study of Race and Ethnicity in America, at Brown University, United States. Her current research focuses on the Chinese diaspora in Latin America and the Caribbean. She recently coedited *Voluntary Organizations in the Chinese Diaspora* (Hong Kong University Press, 2006).

Ben Jiao is a senior researcher and the deputy director of the Contemporary Tibetan Research Institute at the Tibet Academy of Social Sciences, Lhasa.

Rong Ma received his MA and PhD degrees in sociology from Brown University in 1984 and 1987, respectively. He is currently professor of sociology at the Institute of Sociology and Anthropology of Peking University. His major research fields include: ethnic relations, Tibetan studies, education, rural development, and environmental studies. His publications in English include two authored books, *Ethnic Relations in China* (Beijing: China Tibetology Publishing House, 2008) and *Population and Society in*

Contemporary Tibet (Hong Kong University Press, forthcoming). He is also the coeditor of *China's Rural Entrepreneurs* (Times Academic Press, 1995) and *Local Governance and Grassroots Democracy in India and China: Right to Participate* (New Delhi: Sage Publications, 2007).

Xiaoyi Ma, a member of the Hui nationality, earned her PhD in minority education from the Central University for Nationalities in Beijing in 2007. She is currently doing postdoctoral research in the Institute of Education Policy and Law at Beijing Normal University and is a research associate of the Center of Basic Education for Ethnic Minority Studies at the Central University for Nationalities. Her research focuses on education for ethnic minorities in China, education policy, and educational anthropology.

Gerard A. Postiglione is professor and head, Division of Policy, Administration, and Social Science, The Faculty of Education, University of Hong Kong.

Xing Teng is professor of ethnic education studies in the School of Education at the Central University for Nationalities in Beijing and director of the Center of Basic Education for Ethnic Minority Studies. He also serves as vice-secretary-general and member of the Council of the Chinese National Minorities Education Association and counselor of Hong Kong Nursery Fund. His research focuses on educational anthropology, education for ethnic minority in China, bilingual education theory, and bilingual education for ethnic minorities in China, cross-cultural education, and international studies of multicultural education. He has numerous publications on these topics.

Ngawang Tsering is the director of the Contemporary Research Institute at the Tibet Academy of Social Sciences, Lhasa.

Tiezhi Wang, a native Mongolian, holds a PhD in anthropology. He is inspector-general at the PRC State Commission on Ethnic Affairs (SCEA) and adjunct professor at the Central University for Nationalities in Beijing. He has served in various capacities with the PRC SCEA, including as the head of the Division of Higher Education, associate director of the Office of Policy Studies, associate director-general of the Center for Minority Studies, and chief editor of the Ethnic Press in Beijing. He has authored, edited, and coedited over twenty books and published over seventy articles on ethnic affairs in China.

Changjiang Xu is a doctoral candidate at Yunnan University. His research interest is economic anthropology with specific focus on minorities in Yunnan.

Minglang Zhou is associate professor and director of the Chinese program at the University of Maryland, College Park. His research focuses on the sociology of language and ethnic relations in China. He has authored *Multilingualism in China: The Politics of Writing Reforms for Minority Languages 1949–2002* (Mouton de Gruyter, 2003) and edited *Language Policy in the People's Republic of China: Theory and Practice since 1949* (Kluwer Academic Publishers, 2004), *Journal of Asian Pacific Communication* [vol. 16, (2), 2006]*: Special Issue on Language Planning and Varieties of Modern Standard Chinese*, and *Chinese Education and Society* [vol. 41 (6), 2008] on the theme of linguistic diversity and language harmony in contemporary China. He is currently working on his book *Between Integration and Segregation: Models of Nation-State Building and Language Education for Minorities in China,* for which he has been awarded a 2009 American Philosophical Society Sabbatical Fellowship.

Index